PENGUIN BOOKS

UNIVERSAL

Brian Cox is a Professor of Particle Physics and Royal Society University Research Fellow at the University of Manchester, and works at the CERN laboratory in Geneva. He is also a popular presenter on TV and radio.

Jeff Forshaw is Professor of Theoretical Physics at the University of Manchester, specializing in the physics of elementary particles. He was awarded the Institute of Physics Maxwell Medal in 1999 for outstanding contributions to theoretical physics.

UNIVERSAL
A JOURNEY THROUGH THE COSMOS
BRIAN COX & JEFF FORSHAW

PENGUIN BOOKS

PENGUIN BOOKS

UK | USA | Canada | Ireland | Australia
India | New Zealand | South Africa

Penguin Books is part of the Penguin Random House
group of companies whose addresses can be
found at global.penguinrandomhouse.com

First published by Allen Lane 2016
Published in Penguin Books 2017
011

Copyright © Brian Cox and Jeff Forshaw, 2016

The moral right of the authors has been asserted

Designed by Tom Etherington

Set in Sabon and Berthold Akzidenz Grotesk
Printed in Great Britain by Clays Ltd, Elcograf S.p.A.

A CIP catalogue record for this book
is available from the British Library

ISBN: 978–0–241–95317–4

For Brian's dad, David

ACKNOWLEDGEMENTS

For their specific help with various parts of the book, we'd like to thank Richard Battye, Sarah Bridle, Mike Bowman, Bill & Pauline Chamberlain, Ed Copeland, Mrinal Dasgupta, Neal Jackson, Scott Kay, Kevin Kilburn, Peter Millington, Tim O'Brien, Michael Oates, Subir Sarkar, Bob Seymour and Martin Yates. Particular thanks go to Mike Seymour, with whom we have had many enjoyable and helpful discussions.

Special thanks are also due to the team at Penguin and especially Tom Penn, our editor, and Tom Etherington, who produced the figures.

Thanks also to Diane and Sue for their continued guidance and support.

Finally, we would like to thank Peter Saville for his influence on the book, which extends beyond the beautiful cover design.

This book has been a long time in the making and we are deeply grateful for the support and encouragement of our families.

1.
THE STORY
OF THE UNIVERSE

We dare to imagine a time when the entire observable Universe was compressed into a region of space smaller than an atom. And we can do more than just imagine. We can compute. We can compute how hundreds of billions of galaxies emerged from a single subatomic-sized patch of space dwarfed by a mote of dust, and there is precise agreement between those computations and our observations of the cosmos. It seems that human beings can know about the origins of the Universe.

Cosmology is surely the most audacious branch of science. The idea that the Milky Way, our home galaxy of 400 billion stars, was once compressed into a region so vanishingly small is outlandish enough. That the entire visible congregation of billions of galaxies once occupied such a subatomic-sized patch sounds like insanity. But to many cosmologists this claim isn't even mildly controversial.

This is not a book about knowledge handed down from on high. More than anything, it is about how we – all of us – can gain an understanding of the Universe by doing science. You might think that it's impossible for the average person to explore the Universe in much detail: don't we need access to Hubble Space Telescopes and Large Hadron Colliders? The answer is no, not always. Some fundamental questions about our Earth, our Sun, our solar system, and even the Universe beyond, are answerable from your back garden. How old are they? How big are they? How much do they weigh? We will answer these questions by doing science. We will observe, measure and think. One of the great joys of science is to understand something for the first time – to really understand, which is very different from, and far more satisfying than, knowing the facts. We will make our own measurements of the motion of Neptune, follow in the footsteps of the pioneering cosmologist Edwin Hubble in discovering that our Universe is expanding, and make an apparently trivial observation standing on a beach in south Wales.

As the book unfolds, our gaze will inevitably turn out-wards towards the star-filled galaxies. To understand them, we will rely on observations and measurements that we cannot make ourselves. But we can imagine being a part of the teams of astronomers who can. How far away are the stars and galaxies? How big is the Universe? What is it made of? What was it like in the distant past? The answers to these questions will generate a cascade of new ideas, and, before the book is finished, we will be equipped to enquire about the origins of the Universe. Science is an enchanting journey of exploration. It is an exciting, rewarding process and one that leaves scientists with a feeling of being better connected to the world around them. It leaves a sense of awe and humility too; a feeling that the world is beautiful beyond imagination and that we are very privileged to be here to witness it.

Before we begin our journey, however, we will allow our-selves a glimpse of the destination. What follows next is the story of how our Universe evolved from a subatomic patch of space into the oceans of galaxies we see today. Perhaps, by the end of the book, you will judge that it might just be true.

Consider the Universe before the Big Bang. By 'Big Bang' we mean a time 13.8 billion years ago when all the material that makes up the observable Universe came into being in the form of a hot, dense plasma of elementary particles. Before this time, the Universe was very different. It was relatively cold and devoid of particles, and space itself was expanding very rapidly, which means that any particles it may have contained were moving away from each other at high speeds. The average distance between particles was doubling every 10^{-37} seconds. This is a staggering, almost incomprehensible, rate of expansion: two particles one centimetre apart at one instant were separated by 10 billion metres only 4×10^{-36} seconds later; more than twenty times the distance from the Earth to the Moon. We do not know for how long the Universe expanded like this, but it continued for at least 10^{-35} seconds. This pre-Big Bang phase of rapid expansion is known to cosmologists as the epoch of inflation.

Let us focus on a tiny speck of space a billion times smaller than a proton, the atomic nucleus of a hydrogen atom. At first glance, there is nothing particularly special about this tiny patch. It is one small part of a much larger, inflating Universe, and it looks much the same as all the other patches that surround it. The only reason this particular patch deserves our attention is that it is destined, over 13.8 billion years, to grow into our observable Universe: the region of space containing all the galaxies and quasars and black holes and stars and planets and nebulae visible from Earth today. The Universe is far bigger than the observable Universe, but we can't see it all because light can only travel a finite distance in 13.8 billion years.

Before the Big Bang, the Universe was filled with something called the 'inflaton' field; a material thing, like a still ocean filling space. The gravitational effect of the energy stored in the inflaton field caused the Universe's exponential expansion, and this is the origin of its name: it is the field responsible for inflating the Universe. On the whole, the inflaton field remained undisturbed as the Universe expanded, but it was not perfectly uniform. It had tiny ripples in it, as required by the laws of quantum physics.

By the time our observable Universe was the size of a melon, the period of inflation was drawing to a close as the energy driving it drained away. This energy was not lost, however; it was converted into a sea of elementary particles. In an instant, a cold, empty Universe became a hot, dense one. This is how inflation ended and the Big Bang began, delivering a Universe filled with the particles that were destined to evolve into galaxies, stars, planets and people.

We do not currently know which particles were present at the moment of the Big Bang, but we do know that the heaviest particles soon decayed to produce the lighter ones we know today: electrons, quarks, gluons, photons, neutrinos and dark matter.[1] We can also be confident about the particles that populated the Universe when it was around a trillionth of a second old, because we are able to re-create these conditions on Earth, at the Large Hadron Collider.[2] This is the time when empty space became filled with the Higgs field, which caused some of the elementary particles to

---

[1] For a brief resumé on the known elementary particles, see the Appendix.

[2] This is the 27 km circular underground tunnel on the Franco-Swiss border near Geneva, in which protons travelling within a whisker of the speed of light are made to collide. The debris from those collisions tells us how particles interact at energies relevant for studies into the Big Bang.

acquire mass.[3] The weak nuclear force, responsible for the reactions that allow the stars to shine, became distinct from the electromagnetic force at this time.

A millionth of a second after the Big Bang, when the hot plasma had cooled to 10 trillion degrees celsius, the quarks and gluons formed into protons and neutrons, the building blocks of atomic nuclei. Although this primordial Universe consisted of an almost uniform soup of particles, there were slight variations in the density of the soup – an imprint of the quantum-induced ripples in the inflaton field. These variations were the seeds from which the galaxies would later grow.

One minute after the Big Bang, at around a billion degrees, the Universe was cool enough for some of the protons and neutrons to cluster together in pairs to form deuterium nuclei. Most of these then went on to partner with additional protons and neutrons to form helium and, in tiny amounts, lithium. This is the epoch of nucleosynthesis.

For the next 100,000 years or so, little happened as the Universe continued to expand and cool. Towards the end of this time, however, the dark matter gradually began to clump around the seeds sown by the ripples in the inflaton field. Regions of the Universe where there was a slight excess of dark matter grew denser, as their gravity pulled in yet more matter from their surroundings. This is the start of the gravitational clumping of matter that will eventually lead to the formation of galaxies. Meanwhile, photons, electrons and the atomic nuclei bounced and zig-zagged around, hitting

[3] Specifically, in the Standard Model of particle physics, the Higgs field gives mass to the quarks, the electrically charged leptons (the electron is one of these) and the carriers of the weak force (the W and Z bosons). Without the Higgs field, these particles would have zero mass and zip around at the speed of light.

each other so frequently that they formed something resembling a fluid. After 380,000 years, when the observable Universe was a thousand times smaller than it is today, temperatures dropped to those found on the surface of an average sun-like star, cool enough for electrons to be captured in orbit around the electrically charged hydrogen and helium nuclei. Suddenly, across the Universe, the first atoms formed and the Universe underwent a rapid transition from a hot plasma of electrically charged particles to a hot gas of electrically neutral particles. This had dramatic consequences, because photons interact far less with electrically neutral atoms. The Universe became transparent, which means the photons stopped zig-zagging around and started to head off in straight lines. The majority of these photons continued onwards, travelling in straight lines for the next 13.8 billion years. Some of them are just arriving at our Earth today in the form of microwaves. These ancient photons are messengers from the earliest times, and they carry a treasure trove of information that cosmologists have learnt to decode.

As the Universe continued to expand, its denser regions, composed mainly of dark matter, became ever denser under the action of gravity. Hydrogen and helium atoms clustered around the dark matter, and swirling atomic clouds grew until the densest regions collapsed inwards, increasing the pressure and temperature at their core to such an extent that they became nuclear furnaces; the fusion of hydrogen into helium was initiated, and stars formed across the Universe. A hundred million years after the Big Bang, the cosmic dark ages came to an end and the Universe was flooded with starlight. The most massive stars had brief lives and, as they ran out of hydrogen fuel, they began to fuse heavier elements in an ultimately futile battle with gravity: carbon, oxygen, nitrogen, iron – the elements of life – were made this way. When the fuel finally ran out, these stars scattered the newly minted heavy elements across space as they ended their lives as bright planetary nebulae or exploding supernovae. In a final flourish, the violent shock of each exploding supernova synthesized the heaviest elements, including gold and silver. New stars formed from the debris of the old, and congregated in their hundreds of billions in the first galaxies. The galaxies, numbered in hundreds of billions, were moulded into the giant filamentary webs that criss-cross the Universe by the gravitational pull of the dominant dark matter.

4.6 billion years ago in the Milky Way galaxy, a gas cloud enriched in stellar debris collapsed to form our Sun. Shortly afterwards, the Earth formed from the remains of the cloud. Then, 4 billion years ago, in a great ocean created from hydrogen formed in the first minute of the Universe's life and oxygen forged in long-dead stars, the geochemistry of the young Earth became biochemistry: life began. In 1687 Isaac Newton published the *Principia Mathematica*. We've obviously skipped a bit of biology.

This is the broad outline of the story of the evolution of the Universe, from before the Big Bang to Isaac Newton. It seems that collections of atoms on a cooling cinder, in possession of a precious thing called science for barely an instant, have found a way to glimpse the fires of creation. The rest of this book is the story of how we did it.

2.
HOW OLD
ARE THINGS?

The Earth is 4.55 billion years old, give or take 50 million years. This is a figure consistent with independent measurements of the age of the Universe, which place the Big Bang 13.8 billion years ago. It is also consistent with physical biological evidence and our understanding of evolution by natural selection, which suggest that the first living things appeared on Earth around 3.8 billion years ago. The life cycles of stars fit into this timeline too. The age of our Sun is estimated at 4.6 billion years, and similar stars are predicted to live for around 10 billion years before they die. More massive stars have much shorter lifetimes. There must have been time for at least some stars to live and die before the Earth formed, because the Earth is made out of heavy chemical elements like iron, carbon and oxygen: elements that are made inside stars. Leaping forward in time, the basalt columns of the Giant's Causeway in Ireland were formed 60 million years ago, around the time the dinosaurs became extinct. The oldest living tree is a bristlecone pine that lives in the White Mountains in California. It is – as of 2016 – 5066 years old.

All these dates are determined using very different kinds of science, but, remarkably – impressively – they fit together without contradiction. There is nothing special about this particular list; we chose this eclectic bunch simply because they reflect a variety of different 'old' things, and we could have chosen a different list. This raises the question: how do we know how old something is? Age is not a trivial thing to determine, especially for very old things, because it must be inferred indirectly. We can't sit around and watch while the Universe evolves from the hot plasma of its birth. We can't even point to direct evidence for the age of the oldest tree; nobody was around to write about it and record the date when it was a tiny sapling. But we don't need to have been present: knowledge can be acquired indirectly if we do a little detective work to collect evidence and then apply simple logic to draw conclusions. This book is all about taking a scientific approach to securing knowledge of the world around us. This approach is incremental – a framework of knowledge grows over time as we understand more about the Universe – and it sits in stark opposition to haphazard thinking: you don't build a computer by trial and error and you are prone to mistakes if you don't entertain the likelihood that you may be wrong. We trust our lives to scientific knowledge, in hospitals and aeroplanes, and exactly the same type of thinking can be used to great effect elsewhere in our lives. In this book, we will show how far it is possible to travel in understanding the Universe by taking simple, reasoned steps coupled with careful observations. In this chapter, we are going to begin by exploring the science that allows us to measure the age of things with such confidence and precision.

Let's begin with the age of the Earth. A very obvious way to start is to look at what we can see: to ask whether there are any features on the Earth's surface that might give us a clue to its age. To take a careful look at Nature, in other words, and see what we can work out from simple observation. For example, we know that river valleys are cut by flowing water, and that coastlines are subject to erosion. These are features that change with time; therefore, observing them carefully and understanding the physical processes that formed them should allow us to estimate their ages. On larger scales still, could the familiar shapes of the continents and oceans also tell us something about the way they have evolved, and how long it has taken them to do so?

Plate 1 is a map of the Atlantic Ocean and the landmasses that surround it. South America and Africa in particular look as if they fit together. Let's suppose this fit is no accident and make a proposal: the continents were snuggled together at some time in the past, and have been gradually moving apart ever since. If this theory is correct, then we can make a rough estimate of the age of the Atlantic Ocean. Of course, this isn't a new idea – Alfred Wegener's idea of a global-supercontinent that broke up over time as a result of continental drift is over 100 years old. The point here, and throughout this book, is that we can uncover the science for ourselves – we want to follow in the footsteps of the great scientists, to appreciate how irresistible progress comes from simple thoughts. As a first step, we need to confirm that the broad outline of our hypothesis (that South America and Africa were once joined and have been moving apart ever since) is plausible by checking whether the Atlantic is still growing today. If it is, we can measure the current rate of separation of the continents, and – if we make the further assumption that this rate has stayed constant since the time that the continents began to separate – we will be able to make an estimate of the age of the Atlantic. There are a lot of assumptions here, but let's get on with it and see what we find.

If we were very committed experimentalists, we could

measure the movements of the continents ourselves. We could pack a couple of GPS receivers into a rucksack, fly to the eastern coast of Brazil, fix one of the receivers to the ground, fly back across the Atlantic to northwest Africa – a distance of around 4000 km – and set up the second GPS receiver. Over the next few years, we could monitor how the receivers move relative to each other. We don't need to do this, because geologists have already been making such measurements for many years. Quite wonderfully, apart from using GPS receivers, the distance between North America and Europe has also been measured using a pair of radio telescopes (one in Europe and one in the USA) each focused on a distant quasar. Quasars are active galactic nuclei that most probably originate as matter accretes onto super-massive black holes in the centres of galaxies, and they are among the brightest objects and therefore the most distant we can see. Because they are so far away, they serve as excellent fixed points on the sky, which is important for triangulating the distance between Europe and the USA. We describe the measurement in a little more detail in Box 1. Do you remember those school science experiments where you had to begin with the heading 'Apparatus: two large radio telescopes and a grid system comprising active galactic nuclei over a billion light years from Earth'?

Figure 2.2 shows a summary of the results measuring the present-day rates at which the various tectonic plates are moving. It shows that the Atlantic, between northern Brazil and northwest Africa, is currently expanding at a rate of 2.5 cm per year, which is the speed at which fingernails grow.

Working on the assumption that the continents have always been moving apart at this rate, we can now estimate the age of the Atlantic Ocean: 4000 km × 40 years/metre =

BOX 1. MEASURING CONTINENTAL DRIFT   P. 12

The distance between two radio telescopes on the Earth's surface can be determined using a technique known as Very Long Baseline Interferometry. The two telescopes look at the same distant object in the sky, and from the difference in arrival time of light signals – determined using very precise clocks accurate to 1 second in 1 million years – the distance between the telescopes can be determined to millimetre accuracy. Quasars are so bright that they are visible at distances of many billions of light years, and being so far away guarantees that they appear still during the time of the measurement. Over twenty years, telescopes in Westford, Massachusetts, and Wettzell, Germany, have been used to determine the rate at which the Atlantic is opening between Europe and the United States. The data are shown in Figure 2.1, which shows a rate of spreading in this region of 1.7 cm/year. Satellite laser ranging, which involves bouncing laser light off satellites, and GPS measurements are also used along the length of the North and South Atlantic, and give consistent results.

BOX 1. MEASURING CONTINENTAL DRIFT   P. 13

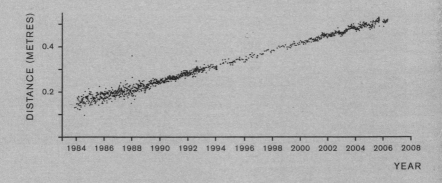

Figure 2.1 The steady rate at which Germany and the USA have been receding from each other in the recent past, as measured by a pair of radio telescopes trained on distant astronomical objects.

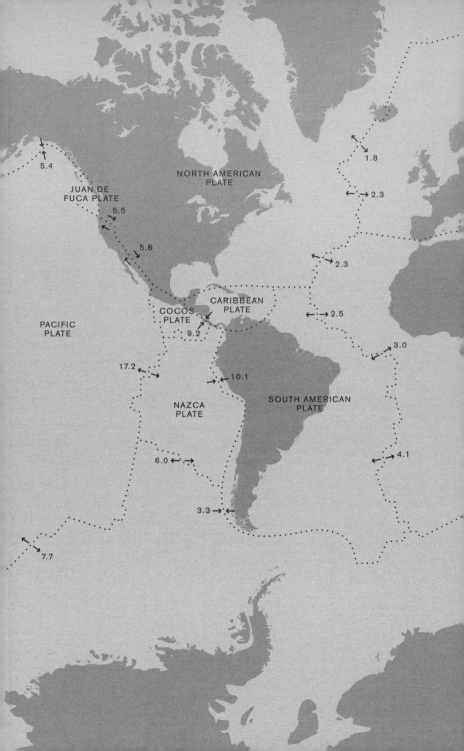

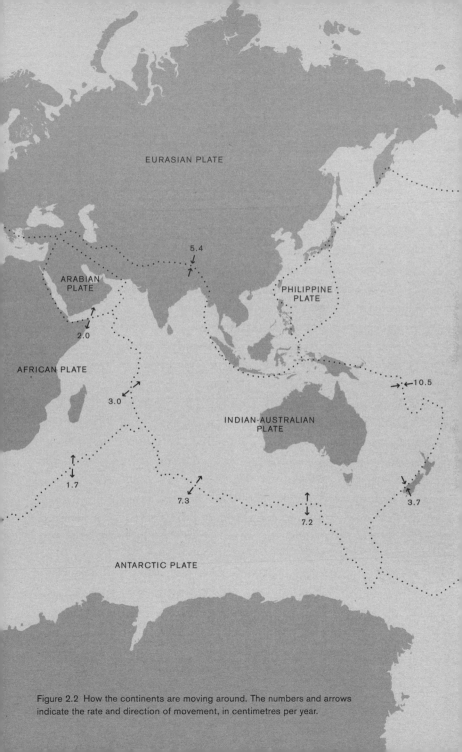

EURASIAN PLATE

ARABIAN PLATE

PHILIPPINE PLATE

5.4

2.0

AFRICAN PLATE

3.0

10.5

INDIAN-AUSTRALIAN PLATE

1.7

7.3

3.7

7.2

ANTARCTIC PLATE

Figure 2.2 How the continents are moving around. The numbers and arrows indicate the rate and direction of movement, in centimetres per year.

160 million years. If this figure is a good estimate, then we now also have a minimum age for the Earth – because obviously it can't be younger than the Atlantic Ocean.

We've just done what could be described as a 'back of the envelope' calculation. Obviously, we'd like to know if our number is anywhere near correct; after all, we did make a bold assertion and a very bold assumption. We asserted that the continents were once part of a single landmass and assumed that they have been moving apart at a steady rate ever since. Let's examine these assumptions more closely and try to judge how reasonable they are.

Look back at the map in Plate 1. It also shows the topology of the Atlantic Ocean's floor. The great range of underwater mountains running down the centre is called the Mid-Atlantic Ridge. This ridge clearly mirrors the shape of the continents on either side; it's also bang in between the two continents, in the middle of the Atlantic, and is currently spewing out material from the Earth's interior: lava that solidifies and forms a crust. This suggests a mechanism that could explain why the continents are continuing to move apart today: new ocean crust is being formed along the Mid-Atlantic Ridge.

All of which seems to indicate that our assertion is in good shape. We could, of course, have been fooled by a series of coincidences: (i) that the coastlines appear to fit together and match the shape of the Mid-Atlantic Ridge; (ii) that the Mid-Atlantic Ridge lies midway between the continents; (iii) that the lava erupting from the Mid-Atlantic Ridge has nothing to do with the currently observed widening of the ocean. But although we can be pretty confident that these are not simply coincidences, nothing we have established so far implies that the separation of these two continents has been proceeding at the same rate for over a hundred million years, and we must admit that, at this stage, this assumption is a blind guess.

So let's bring in some serious science. For decades, geoscientists have meticulously examined ocean floors across the globe and determined the age of the rocks on the seabed. This is a difficult task, and requires some beautiful science that we will discuss in a moment (see also Box 2). For now, let us just present the data, which is shown in Plate 2. There is a very clear pattern in the Atlantic: the youngest rocks lie along the Mid-Atlantic Ridge; the oldest are to be found bordering the continents. This fits very nicely with our proposal that the Atlantic was formed by sea-floor spreading from the Mid-Atlantic Ridge; if we are right, the rocks on the seabed should indeed get progressively older the further we travel from the ridge. Notice also that there are no sharp transitions where the rocks suddenly get much older, nor are there any extended regions where the rocks are all the same age. This is what we would expect if the rate at which new rock is being formed along the Mid-Atlantic Ridge has remained roughly constant during the time that the continents have been moving apart. The final observation we can make is to look at the age of the rocks lining the ocean floors along the edges of the continents. These are dated to be around 180 million years old – in broad agreement with our back-of-the-envelope calculation.

We haven't yet described how we go about dating rocks directly. But we can say that our suggestion that the Atlantic Ocean was created by the continents drifting apart due to geological activity along the Mid-Atlantic Ridge is consistent

BOX 2. SEA-FLOOR SPREADING   P. 18

The age of sea-floor rocks in the Atlantic is determined by
exploiting the fact that sea-floor basalt is magnetized in a
stripy pattern, as shown in Figure 2.3. The stripes, typically
some tens of kilometres wide, are formed as new rock
spews out of the ridge and becomes magnetized by the
Earth's magnetic field. When the rock freezes, the magnetic
orientation gets frozen within it. The stripes appear because
the Earth's magnetic field flips its direction from time to time,
and these changes of direction are encoded in the rocks.
We can therefore map the temporal evolution of the sea-floor
by measuring the barcode-like pattern of stripes, known as
'polarity chrons', so long as we have some method of setting
the timescale for the flips in the Earth's magnetic field. And
we do: radiometric dating methods have been used to date
rocks in other locations, such as on-land lava flows. The
barcode patterns match.

In December 1968 and January 1969, the *Glomar
Challenger*, a scientific research drill-ship, acquired a very
important dataset by drilling a series of seventeen holes in
the equatorial and South Atlantic, many of them traversing
the Mid-Atlantic Ridge. The samples the *Glomar Challenger*
collected were dated mainly by paleontological methods,
which involved looking for tiny fossils in the sample cores and
matching them with known stages in the evolution in the flora
and fauna of the oceans (whose ages themselves are fixed
using radiometric methods). The shipboard scientists involved
analysed the cores and found an age–distance relationship
from the Mid-Atlantic Ridge that is remarkably consistent with
the assumption that the sea-floor has been spreading at a
constant rate. They found sediments sitting directly above the
sea-floor with ages ranging from 10 million years for samples
200 km from the ridge, all the way to 70 million years for
samples taken 1300 km from the ridge, corresponding to a
sea-floor spreading rate of close to 2 cm/year.

BOX 2. SEA-FLOOR SPREADING   P. 19

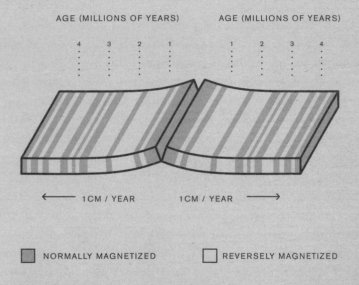

Figure 2.3 The ages of rocks forming along a rift valley, such as the one running along the Mid-Atlantic Ridge. The barcode stripes are very distinctive and are due to the fact that the Earth flips its magnetic field around every so often.

with the measured age of the rocks on the ocean floor.

Logical consistency and the accumulation of evidence are very important features of the way modern science works. Consider, for example, what would happen to our previous logic if the Atlantic were significantly younger than 160 million years. For the sake of argument, let's go with Bishop James Ussher, and say it is around 10,000 years old. This rather casual level of precision is doing the good bishop a disservice, because he was very specific. He asserted that the world was created on the evening of 22 October 4004 BC. The bishop performed his calculations in the late seventeenth century, using historical records and the Bible. We, on the other hand, are operating on the back of an envelope, which means we are content to work with round numbers.

If we want to accommodate an Atlantic that is 10,000 years old, but still accept that the two continents were both close together at some point, then the rate at which the continents moved apart would have had to have been much faster than the currently observed 2.5 cm per year. Instead, we would require an expansion rate of the order of 400 metres per year for most of the 10,000-year period.

The problem with an expansion rate of 400 metres per year is that the rocks along the Atlantic shores are measured to be 180,000 years old, a date that is in good agreement with the 2.5 cm per year spreading rate. If we insisted on a 10,000-year-old Earth, then it must follow that the rock ages are wrong by precisely the same factor as the spreading-rate estimate is wrong. This would be quite a coincidence.

With Bishop Ussher's dating still in mind, a second possibility might be that the continents were never in fact close together, but instead they were originally created 4000 km apart, 10,000 years ago. In that scenario, the fact that the observed drift rate of 2.5 cm per year just happens to be consistent with the age inferred from dating the rocks must be regarded as a meaningless coincidence, not least because we would also need to reject as wrong the methods used to date the rocks. In addition, we'd also have to suppose that the two continents and the Mid-Atlantic Ridge all fit together quite by accident. It is clear then that the case for a young Earth requires we reject the most obvious interpretation of the facts and appeal instead to coincidence and error. We have only been studying the case of the Atlantic Ocean so far and we will meet some more examples of very old things in due course. It is up to you to judge the extent to which the evidence is convincing.

The reason that it is so difficult to make an argument against the Atlantic Ocean being around 160 million years old is that independent measurements, relying on completely different science, combine to provide a consistent picture of what happened. It is easy to cook up a scenario, however fanciful, that casts doubt on some measurement or other. But it is usually extremely difficult to argue for a radical change in one area without making large parts of the whole inter-linked edifice inconsistent. Given that the scientific edifice is also the thing that keeps your lights on, keeps aircraft in the sky and makes your computer work, this is not usually a sustainable position to take. Our modern scientific world-view is a universal one, and this is a key reason why it is so robust and successful.

One of the most precise ways of dating old rocks is through radiometric techniques. The key idea is that certain types of atoms are radioactive, which means that they can spontaneously transform into atoms of a different type. (In Box 3 (pp. 39–48), we provide a primer on the basics on atoms and radioactivity.) This transformation process is known as radioactive decay. If we know the rate at which a particular type of atom decays, then by counting the number of those atoms present in a rock sample we can obtain a measure of how much time has passed since it was formed. We do not need to know what causes atoms to decay (for that we have to understand some quantum physics); we just need to know the rate at which atoms decay, which is called the half-life. The half-life expresses how long, on average, it takes for half the atoms in a sample to decay. For example, if we know that a rock sample initially contained N radioactive atoms, and we measure that it currently contains $N/4$ atoms – that's to say, a quarter of the radioactive atoms originally present – we can deduce immediately that two half-lives have elapsed since the rock was formed.

Some atoms have short half-lives of much less than one second; others have long half-lives, reaching into the billions of years. If we want to determine the age of a rock, the best way would be to count atoms whose half-life is not too different from the age of that rock. If the half-life is much less than the age, most of the radioactive atoms will have decayed away, and we will have a difficult job counting the small number that remain. If the half-life is much greater than the age, very few atoms will have decayed and we might struggle to determine any significant deviation from the initial number. None of this would be of any practical use if radioactive atoms were rarely found in rocks. Fortunately, they are relatively common.

You may have already noticed that there could be a flaw in our plan. How could we possibly know how many radioactive atoms were present in a given rock when it first formed? This might seem to scupper the half-life dating procedure. However, there is a beautiful way to sidestep the problem, known as the isochron method.

In order to understand the isochron method, let's look at a specific atom, rubidium-87, which we will write as ^{87}Rb. Rubidium is chemically similar to potassium, about as abundant as zinc, and often found in common, potassium-rich minerals in rocks. It is radioactive, with a very long half-life of 48 billion years. Being barely radioactive is a bonus for dating the oldest rocks on Earth because, as we will see, they have ages of several billion years. When a rubidium atom decays, it converts into an atom of strontium-87 (^{87}Sr). We can count the numbers of ^{87}Rb and ^{87}Sr atoms in a sample of rock. The clever part of the isochron method is to exploit the existence of a different sort of strontium atom, strontium-86. ^{87}Sr and ^{86}Sr are different isotopes of strontium; the only difference is that ^{86}Sr has 1 less neutron in its nucleus. This means that they are chemically identical – and this is crucially important. Also crucially, ^{86}Sr cannot be produced through radioactive decay, which means that any ^{86}Sr now present in the rock sample was there when it was originally formed.

In a sample from the rock we want to date, we count the number of ^{87}Rb atoms and divide by the number of ^{86}Sr atoms. We also count the number of ^{87}Sr atoms and divide that by the number of ^{86}Sr atoms. We mark these two ratios as a point on a graph, as shown in Figure 2.4. We repeat this counting process for several different samples, each taken

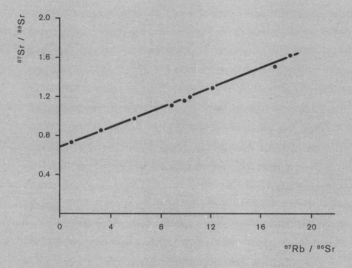

Figure 2.4 Two isochron plots used to date the ages of rocks. The right-hand plot is for samples taken from the chondrite meteorite Tieschitz that fell in what is now the Czech Republic in 1878. The left-hand plot is for samples taken from Isua in Greenland.

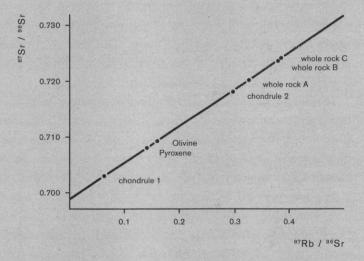

from the rock we want to date. The different samples might be chunks of rock taken from a large bed, or they might be samples of different minerals taken from the same piece of rock. If the samples are all absolutely identical to each other, then we will find the same ^{87}Rb to ^{86}Sr ratios in all of them, and this will simply generate lots of points on top of each other on the graph. But if the different samples had differing initial amounts of ^{87}Rb, then the ratios will be different and we will get several points on our graph.

The striking thing about the graphs shown in Figure 2.4 – which contain real data – is that the different points, corresponding to different initial concentrations of rubidium atoms in each sample, lie on a straight line. This is not a coincidence; in fact, it's the clever bit.

To understand what is happening here, imagine that we make an isochron plot immediately after the rock has formed from a molten state. Since the two isotopes of strontium have identical chemical properties, the ratio $^{87}Sr/^{86}Sr$ will be the same initially in every sample. For example, if there are 7 ^{87}Sr atoms for every 10 ^{86}Sr atoms throughout the initial molten mix, then this ratio will be preserved in every mineral within every newly solidified rock sample. This is because there is no reason for any particular mineral to form using one strontium isotope over the other, since the two strontium isotopes are chemically identical. This means that at time-zero, just after the rock forms, the isochron plot will be a horizontal straight line. This is illustrated in Figure 2.5. The Rb/Sr ratio will vary because rubidium and strontium are chemically different from each other. If one sample is taken from a potassium-rich mineral, for example, it is likely to have more rubidium than a sample taken from a mineral with less

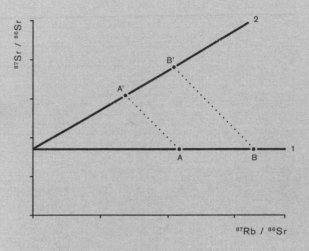

1. Rock just formed
2. Same rock much later on

Figure 2.5 Illustrating how the line on an isochron plot tilts away from the horizontal as time passes. Point A moves to A' and B moves to B'.

potassium, because rubidium behaves like a potassium substitute. We therefore expect the initial Rb/Sr ratio to vary
with the potassium content of the different mineral samples.

As time advances, rubidium atoms decay; as a result, the
amount of ^{87}Sr inside the rock starts to rise. This means that
a point on the original plot moves to the left by an amount
proportional to the number of rubidium atoms that have
decayed. The point also moves upwards by exactly the same
amount, because for every decaying Rb atom a new atom of
^{87}Sr is created. This happens for all the original points, and
results in the line tilting from the horizontal, as illustrated in
Figure 2.5. As more time passes since the rock was formed,
the line tilts further. If there was no initial rubidium present
in the sample then there can never be any new strontium
generated, so the left-hand point on the line must stay fixed,
as indicated in the figure. Crucially, the points should remain
on a straight line. The line simply tilts with time, and the tilt
from the horizontal is what tells us the age of the sample.

This is how the ingenious isochron method works. We
never needed to know the initial concentrations of any of the
atoms; we simply need a set of different samples, with different initial concentrations of rubidium and strontium.

We'll show what we mean. Suppose we happen to measure
the rock's age after 1 half-life of rubidium – which is admittedly difficult, because that's longer than the age of the
Universe, but we are going for simple mathematics. In this
case, a point on the initial isochron moves halfway to the
left, because there are half as many Rb atoms as there were
initially. The point also moves upwards by exactly the same
amount (because an equal number of new ^{87}Sr atoms are
created). This corresponds to a tilt of 45 degrees to the hori-

zontal. An isochron tilted at 45 degrees therefore means that one half-life has elapsed.

For those who want a bit more detail: if g is the fraction of the original rubidium that has decayed since the rock was formed, then a point on the initial horizontal isochron moves to the left a distance g times its initial value and moves up by the same distance. This means the tangent of the tilt relative to the horizontal is $g/(1-g)$. This is very nice, because g depends only upon the half-life and the time since the sample was formed. Measuring g from the slope of the graph tells us the age of the sample without us needing to know how much rubidium or strontium was initially present in it. For the case of the meteorite illustrated in Figure 2.4, the tangent of the tilt is approximately $(0.7325 - 0.699)/0.5 = 0.067$. This implies g is 0.063, which means there have been m half-lives where $(1/2)^m = 1 - 0.063 = 0.937$, which tells us that $m = 0.094$. Since the half-life of rubidium is 48 billion years, this particular rock (multiplying 48 billion by 0.094) is dated at 4.5 billion years old.

The isochron method represents a beautifully simple piece of science. It tells us how much time has passed since the isochron was horizontal, which is the time the rock cooled and turned solid. In order for the points on the isochron to fall on a straight line it must be the case that for every Rb atom that decays in the rock a new ^{87}Sr atom appears. Moreover, the rock has to have acted like a sealed capsule, so that no Rb or Sr atoms entered or left it following its original formation. If the rock has not behaved like a sealed capsule then the points will not line up in a straight line. This is nice, because it allows us to check whether the rock interacted with its environment in some way, which is an indication that the dating will be unreliable. The flip side is that if the data points do all line up we can be supremely confident that we are measuring the time since the rock was last in a molten state. The isochron method is self-checking, which adds greatly to the confidence we can have in its results.

The North Atlantic craton is an ancient part of the Earth's crust that is exposed in Greenland, the coast of Labrador in northern Canada and parts of northwest Scotland. It is primarily composed of granitoid gneiss, a type of metamorphic rock, which means it was formed from other rocks under very high temperatures and pressures. The left-hand plot in Figure 2.4 shows a rubidium–strontium isochron for samples taken from the North Atlantic craton in Isua, on the southwest coast of Greenland. The data points lie on a straight line, which tells us that the rock has not been altered significantly since its formation. The gradient of the line dates the rock samples to a common age of 3.66 billion years, with an uncertainty of 0.06 billion years. If you followed through the calculation above you will be able to get this age for yourself.

Many other samples have been dated across the North Atlantic craton, sometimes using different radioactive atoms and their isochrons. The measured ages vary mainly between 2.6 and 3 billion years, but the most ancient are around 3.8 billion years old. These ages are consistent with the most ancient rocks found at sites elsewhere in the world. The oldest known rocks are found in the Slave craton, in northwest Canada, and are close to 4 billion years old. Zircon, a mineral that contains small amounts of radioactive uranium and thorium, can be dated by several methods, including a uranium–lead isochron. Zircon grains have been found in

Jack Hills in Western Australia that are 4.404 billion years old, with an uncertainty of 0.008 billion years. These are the oldest materials that have been found on Earth.

No intact crust has been found that is older than 4 billion years. This may be due to the fact that the hot, young Earth vigorously reprocessed its crust, thereby resetting the radiometric clocks in the rocks. Perhaps this reprocessing was aided by a heavy bombardment of meteorites – or, possibly, older rocks do exist and we haven't found them yet. Zircon grains can remain intact in more extreme conditions than most rocks, which is probably why they have been discovered with significantly older isochron-derived ages. In any case, the radiometric dating of rocks and zircon grains allows us to claim that the Earth is at least 4.4 billion years old.

But what exactly do we mean by 'the age of the Earth'? Presumably it did not form in an instant. We can avoid the difficult question of the Earth's birthday if the planetary formation process was not too lengthy compared to its age. Theoretical modelling of planetary formation suggests that this is the case, and that a mere 0.1 billion years is the absolute upper limit on the time it would have taken for the Earth to form from the solar nebula. Going further, we might suppose that all of the planets in the solar system are approximately the same age, because they all formed from the same nebula. This is something we can test, because we have many rock samples from space, in the form of meteorites, and four tonnes of Moon rocks returned by the Apollo astronauts. Many of these samples have been dated using isochron methods.

The youngest Moon rocks are found to be 3.2 billion years

old; the oldest are 4.5 billion years old. Twelve samples have ages older than 4.2 billion years, with an uncertainty of around 0.1 billion years. The geology of these rocks is consistent with them having formed as the lunar crust cooled, and indicate that the Moon is 4.53 billion years old. The radiometric dating of almost a hundred meteorites, such as the Tieschitz meteorite we met above, reveals that the vast majority formed between 4.4 and 4.6 billion years ago, while the younger ones show clear evidence of severe shock-heating and metamorphism, which will have reset the radiometric clocks. Bringing all the data together, including dates from another precision technique called lead-isotopic dating, the current best measurement of the age of the Earth is 4.55 billion years, with an uncertainty of 0.02 billion years.

We could stop here, satisfied that we have a precision measurement of the age of the Earth gleaned from material taken from different sites, and cross-checked using rocks from the Moon and meteorites that have been wandering through the solar system untouched and uncontaminated since their formation, using a variety of radiometric dating methods. In the spirit of this book, however, we should ask if there is some other, completely independent, measurement that does not rely on radioactive decay. And, wonderfully, there is such a method. We can measure the age of the Sun.

The Sun shines by releasing energy through the nuclear fusion of hydrogen into helium. Fusion also took place in the first few minutes in the life of the Universe, creating the primordial helium that constitutes a quarter of the visible matter present in the Universe today (the remaining three quarters being hydrogen). When the Sun formed, its initial composition roughly reflected this universal ratio of hydrogen to helium, but, over time, the fraction of helium in the Sun increased as the Sun burned. If we can measure the amount of helium in the Sun that has been produced by fusion reactions in its core, and if we know the rate of those reactions, then we can estimate the age of the Sun by calculating how long it would take to make that amount of helium.

This sounds like a tall order. We can hardly go to the Sun, dig into its core and take samples. (As we'll see in Chapter 3, it is challenge enough to work out the composition of the Earth.) Remarkably, however, studies of the Sun over the past fifty years have allowed scientists to investigate its interior, using a technique known as helioseismology. Pressure waves in the Sun cause it to ring like a bell, and analysing the way that it vibrates allows for its composition to be deduced.[1] The method is not unlike the way you might ascertain whether something is hollow by knocking it. These studies indicate that around 4.2% of the Sun's total mass is helium that has been produced as a result of nuclear fusion. Figure 2.6 shows how the helium abundance changes with distance from the centre of the Sun. You can see that it increases towards the core and that it levels off at around 27%, which is close to the fraction we expect to be present as a result of helium formation shortly after the Big Bang (we'll have more to say about Big Bang Nucleosynthesis in Chapter 6). The dark shaded region in the figure, at radii smaller than about 0.2, is the helium produced through fusion.

---

[1] This is done by observing how the solar spectrum, which we will meet in the next chapter, changes over time periods of several minutes.

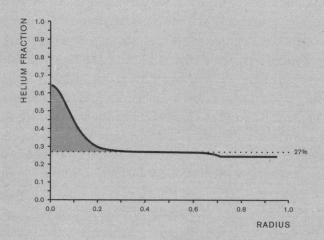

Figure 2.6 The helium fraction in the Sun (by mass). There is a clear excess in the core, i.e. for radii less than 20% of the Sun's radius. This is the result of hydrogen fusing into helium. The residual level at 27% is the helium that was present in the swirling gas out of which the Sun formed.

We need now to work out how long it would take for the Sun to make that much helium. We can estimate this because we know the Sun's total energy output, and we know how much energy is released by the fusion of four hydrogen atoms into one helium atom. Let's deal with each of these in turn.

The Sun radiates energy at a rate of 3.9×10^{26} watts, a number first measured to a reasonable accuracy in 1838 by the French physicist Claude Pouillet. It's relatively simple to do: you could make an estimate yourself using some water, a thermometer, a watch and a bucket. The basic idea is to measure how long it takes direct sunlight falling on the Earth to raise the temperature of a known volume of water, with a known surface area, by 1 degree. This enables you to estimate how much solar energy per square metre per second arrives at the Earth's position in space, 93 million miles away from the Sun. This quantity is known as the solar constant. Modern measurements of the solar constant are made by satellites above the Earth's atmosphere, but you'll get it to within 10% or so if you perform the measurements at the top of a mountain and are careful with your mathematics and your bucket – as Pouillet was. The solar constant is approximately 1.36 kilowatts per square metre, and varies by about 0.1% with changes in solar activity and by about 7% over the course of a year, due to the eccentricity of the Earth's orbit. The total solar energy output can then be calculated, if you know the distance from the Earth to the Sun. (We'll see how this can be measured in Chapter 3.)

The details of the fusion chain that leads to helium production in the Sun are illustrated in Figure 2.7. It's quite complicated, but the final answer is simple. Four hydrogen nuclei are consumed to make one helium nucleus, and in the

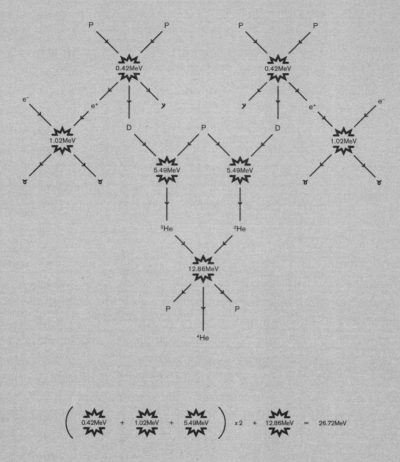

Figure 2.7 The reactions that dominate helium-4 production in the Sun. The fusion of two protons to produce a deuteron yields 0.42 MeV. The emitted positron annihilates with a nearby electron to liberate a further 1.02 MeV. And the fusion of the deuteron with another proton to produce helium-3 yields 5.49 MeV. Two of these chains yield a total of 2 × (0.42 + 1.02 + 5.49) = 13.86 MeV. Finally, the two helium-3 nuclei fuse to produce the helium-4 nucleus, with a further 12.86 MeV of energy released. The net effect of all of these nuclear fusion reactions is for four protons to convert into one helium nucleus with the release of 26.72 MeV in energy.

process 26.7 MeV of energy is released,[2] of which 98% is radiated away as light. The missing 2% of the energy is carried away by neutrinos.

Because the Sun is releasing energy at a rate of 3.9×10^{26} watts, we can deduce that it is making $3.9 \times 10^{26}/(26.72 \times 1.6 \times 10^{-13}) = 9.1 \times 10^{37}$ helium nuclei per second. If we suppose that the Sun has been doing this ever since it was born, then we can deduce that the age of the Sun is equal to the total mass of helium produced by fusion divided by the mass of helium produced every second, which is $4.2\% \times 1.99 \times 10^{30}$ kg / $(6.64 \times 10^{-27}$ kg $\times 9.1 \times 10^{37})$ seconds. We have taken 1.99×10^{30} kg for the mass of the Sun and 6.64×10^{-27} kg for the mass of one helium nucleus. This gives us an age of 1.4×10^{17} seconds, which is equal to 4.4 billion years.[3] Amazing!

We are surely allowed an exclamation mark. We have set great store on achieving consistency between scientific results. Here we have presented a calculation for the age of the Sun that uses our understanding of nuclear fusion reactions, a measurement of the amount of solar energy falling on the Earth every second, the distance of the Earth to the Sun and measurements of the amount of helium in the Sun. This calculation gives an age consistent with the age of the oldest rocks on Earth and on the Moon, and the ages of meteorites that have fallen on the Earth. These were calculated using our understanding of the radioactive decay of heavy atoms, which is a completely different physical process.

[2] MeV is a convenient unit of energy used in atomic and nuclear physics. For a primer on units, see the Appendix.

[3] More precise calculations lead to an age of around 4.6 billion years.

Remarkably, we have found that the ages of the Sun and Earth are very nearly the same, which fits perfectly with the idea that they were formed out of the same cloud of collapsing gas and dust around 4.6 billion years ago.

BOX 3. EVERYTHING IS MADE OF ATOMS   P. 39

If, in some cataclysm, all scientific knowledge were to be destroyed, and only one sentence passed on to the next generations of creatures, what statement would contain the most information in the fewest words? I believe it is the atomic hypothesis (or the atomic fact, or whatever you wish to call it) that *all things are made of atoms – little particles that move around in perpetual motion, attracting each other when they are a little distance apart, but repelling upon being squeezed into one another.* In that one sentence, you will see, there is an enormous amount of information about the world, if just a little imagination and thinking are applied.

(Richard Feynman, *Six Easy Pieces*, p. 4)

The child-like simplicity of the question 'What if I divide this thing in half ... and in half again ... and again ...?' belies its sophistication. The answer to this question is not fully known, and it leads us downwards into the world of atoms: a world in which tiny particles dance around according to the crazy rules of quantum physics. We do not need to delve into those rules here, instead we'd like to get a rough-and-ready appreciation of some basic properties of everyday 'stuff'. Let's start out with a summary of what we know.

   We know that all ordinary matter is made up of atoms, from the Sun to a lump of rock to the air we breathe, and that there are 98 different types of atom that occur naturally on the Earth (humans have managed to build a further 20 different types). These are all listed in the periodic table, shown in Figure 2.8. The most commonly occurring type of atom in the Universe is hydrogen; over 90% of the atoms in the Sun are hydrogen (the remainder being almost entirely helium). On Earth, the atomic composition is more diverse. The oceans are mainly hydrogen and oxygen, while the most abundant atoms in the Earth's crust are oxygen and silicon. In the atmosphere they are nitrogen and oxygen, while the core is predominantly made of iron.

   We know that each atom is made up of a central nucleus, which contains almost all of the mass of the atom, and surrounding it is a 'swarm' of tiny electrons (tiny in

BOX 3. EVERYTHING IS MADE OF ATOMS P. 40

1.00794 1																
H Hydrogen																

6.941 3	9.012182 4
Li Lithium	**Be** Beryllium

22.98976 11	24.3050 12
Na Sodium	**Mg** Magnesium

39.0983 19	40.078 20	44.95591 21	47.867 22	50.9415 23	51.9962 24	54.93804 25	55.845 26	58.93319 27
K Potassium	**Ca** Calcium	**Sc** Scandium	**Ti** Titanium	**V** Vanadium	**Cr** Chromium	**Mn** Manganese	**Fe** Iron	**Co** Colbalt
85.4678 37	87.62 38	88.90585 39	91.224 40	92.90638 41	95.69 42	(98) 43	101.07 44	102.9055 45
Rb Rubidium	**Sr** Strontium	**Y** Yttrium	**Zr** Zirconium	**Nb** Niobium	**Mo** Molybdenum	**Tc** Technetium	**Ru** Ruthenium	**Rh** Rhodium
132.9054 55	137.327 56	174.9668 71	178.49 72	180.9478 73	183.84 74	186.207 75	190.23 76	192.217 77
Cs Caesium	**Ba** Barium	**Lu** Lutetium	**Hf** Hafnium	**Ta** Tantalum	**W** Tungsten	**Re** Rhenium	**Os** Osmium	**Ir** Iridium
(223) 87	(226) 88	(262) 103	(261) 104	(262) 105	(266) 106	(264) 107	(277) 108	(268) 109
Fr Francium	**Ra** Radium	**Lr** Lawrencium	**Rf** Rutherfordium	**Db** Dubnium	**Sg** Seaborgium	**Bh** Bohrium	**Hs** Hassium	**Mt** Meitnerium

138.9054 57	140.116 58	140.9076 59	144.242 60	(145) 61	150.36 62
La Lanthanum	**Ce** Cerium	**Pr** Praseodymium	**Nd** Neodymium	**Pm** Promethium	**Sm** Samarium
(227) 89	232.0380 90	231.0358 91	238.0289 92	(237) 93	(244) 94
Ac Actinium	**Th** Thorium	**Pa** Protactinium	**U** Uranium	**Np** Neptunium	**Pu** Plutonium

Figure 2.8 The periodic table of the elements.

BOX 3. EVERYTHING IS MADE OF ATOMS   P. 41

4.002602 2
He
Helium

10.811 5	12.0107 6	14.0067 7	15.9994 8	18.998403 9	20.1797 10
B	C	N	O	F	Ne
Boron	Carbon	Nitrogen	Oxygen	Fluorine	Neon

26.98153 13	28.0855 14	30.97696 15	32.065 16	35.453 17	39.948 18
Al	Si	P	S	Cl	Ar
Aluminium	Silicon	Phosphorus	Sulfur	Chlorine	Argon

58.6934 28	63.546 29	65.38 30	69.723 31	72.64 32	74.92160 33	78.96 34	79.904 35	83.798 36
Ni	Cu	Zn	Ga	Ge	As	Se	Br	Kr
Nickel	Copper	Zinc	Gallium	Germanium	Arsenic	Selenium	Bromine	Krypton

106.42 46	107.8682 47	112.441 48	114.818 49	118.710 50	121.760 51	127.60 52	126.9044 53	131.293 54
Pd	Ag	Cd	In	Sn	Sb	Te	I	Xe
Palladium	Silver	Cadmium	Indium	Tin	Antimony	Tellurium	Iodine	Xenon

195.084 78	196.9665 79	200.59 80	204.3833 81	207.2 82	208.9804 83	(210) 84	(210) 85	(220) 86
Pt	Au	Hg	Tl	Pb	Bi	Po	At	Rn
Platinum	Gold	Mercury	Thallium	Lead	Bismuth	Polonium	Astatine	Radon

(271) 110	(272) 111	(285) 112	(286) 113	(289) 114	(288) 115	(292) 116	117	(294) 118
Ds	Rg	Cn	Uut	Fl	Uup	Lv	Uus	Uuo
Darmstadium	Roengenium	Copernicium	Ununtrium	Flerovium	Ununpentium	Livermorium	Ununseptium	Ununoctium

151.964 63	157.25 64	158.9253 65	162.500 66	164.9303 67	167.259 68	169.9342 69	173.054 70
Eu	Gd	Tb	Dy	Ho	Er	Tm	Yb
Europium	Gadolinium	Terbium	Dysprosium	Holmium	Erbium	Thulium	Ytterbium

(243) 95	(247) 96	(247) 97	(251) 98	(252) 99	(257) 100	(258) 101	(259) 102
Am	Cm	Bk	Cf	Es	Fm	Md	No
Americium	Curium	Berkelium	Californium	Einsteinium	Fermium	Mendelevium	Nobelium

BOX 3. EVERYTHING IS MADE OF ATOMS   P. 42

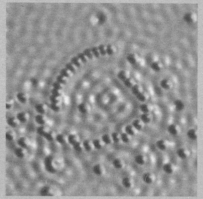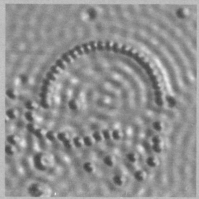

comparison to the nucleus). The nucleus is typically just a
few femtometres across, and the electrons dance around
it: infinitesimal specks at distances of a few tenths of a
nanometre across. To give an idea of what this means, if we
were to scale up an atom so that the nucleus was the size of
a pea, the electrons would be like tiny grains of sand hopping
around a kilometre away. In other words, an atom is almost
entirely empty space. In Box 4 (pp. 49–51) we show how you
can make an estimate of the size of an atom yourself.

But these tiny, point-like electrons, dancing at great
distances in relation to the nucleus, are anything but
ephemeral. For it is their dancing motion, described by
the rules of quantum theory, that determines the way that
collections of atoms communicate with each other; in other
words, the electrons are what fixes the chemistry.

As far as the chemistry is concerned, the nucleus is
an inert object and acts only as a heavy source of positive
electric charge, which is responsible for holding the
negatively charged electrons in a kind of orbit around it. In
everyday life, the electrons do the hard work of negotiating

BOX 3. EVERYTHING IS MADE OF ATOMS   P. 43

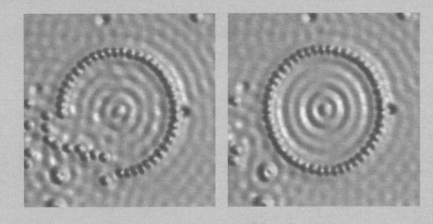

Figure 2.9 A ring of 48 iron atoms adsorbed onto a copper surface. The
pictures are taken with a scanning tunnelling microscope and the circular
waves trapped inside the 'corral' correspond to waves of electron density. It
shows how electrons behave like waves, which is a feature of quantum theory.

how the atoms and molecules behave, while the nuclei sit
inert at the heart of atoms, shielded from the maelstrom of
electron activity.

The nuclei are far from boring, though. Breaking apart
heavy nuclei (fission) or fusing light nuclei together can be
used to generate vast amounts of energy. Fission is what
underpins the reactors that are used throughout the world
today, and fusion reactors promise to deliver a clean and
virtually limitless supply of energy. Looking inside nuclei is
also mandated by our natural human curiosity – we want to
know what they are made of. So far, we know that nuclei
are made of protons and neutrons and that they, in turn, are
built from quarks and gluons. The story seems to stop at
quarks, gluons and electrons because the Standard Model

BOX 3. EVERYTHING IS MADE OF ATOMS   P. 44

of particle physics describes these objects without any need for substructure. In other words, it might not make any sense to ask 'What happens if I chop an electron in half?' or 'What is an electron made from?' The fact that the 'What happens if I chop this thing in half' sequence may eventually come to an end is something that is reasonable in quantum physics, not least because the more we try to pin down the location of very tiny objects the more elusive they become. So while quantum physics does not exclude the possibility that particles like electrons and quarks have substructures, nor does it demand it.

Nuclei are also interesting because they can perform acts of alchemy. By that we mean that a nucleus of one type can spontaneously change into a nucleus of a different type. For example, in nuclear alpha decay a nucleus can eject an 'alpha particle', which is in fact the nucleus of a helium atom. Certain nuclei have a propensity to eject alpha particles: uranium and thorium produce most of the helium on Earth this way. The alpha emitter plutonium-238 is used to power heart pacemakers, and americium-241 (which is made inside nuclear reactors and is a by-product of the Manhattan project) is used in smoke detectors. The americium produces alpha particles that collide with air molecules, knocking electrons off them to produce an electric current. The current falls if smoke particles enter the detector and prevent the alpha particles from ionizing the air, which in turn triggers the alarm. Alpha emission was a total mystery prior to the arrival of quantum theory – not least because it is the subatomic equivalent of throwing a tennis ball at a brick wall and occasionally seeing it pass through. By which we mean that the alpha particle manages to escape from inside the nucleus even though it ought to be trapped within it, just as a tennis ball ought to be trapped on one side of a brick wall. This weird effect is called quantum tunnelling.

Also extremely puzzling is the simple fact that nobody can predict exactly when a particular atom will undergo a radioactive decay, although we can say how long it will take on average. For example, we might say that there is a 50% chance of an atom transmuting in a certain interval of

BOX 3. EVERYTHING IS MADE OF ATOMS   P. 45

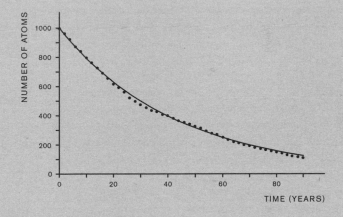

Figure 2.10 The number of atoms in a hypothetical sample of atoms whose half-life is 30 years. The dots are for a specific sample while the curve is what you would get if you averaged over many samples.

BOX 3. EVERYTHING IS MADE OF ATOMS   P. 46

time, called the half-life. This is illustrated in Figure 2.10, which shows how the number of radioactive caesium-137 atoms might change over time. Caesium-137 decays to barium-137 when one of its neutrons turns into a proton, with the concurrent emission of an electron and an electron-antineutrino.[1] This type of decay is known as beta decay and it is a consequence of the weak nuclear force. Figure 2.10 shows that, of a hypothetical initial sample of 1000 atoms, 474 remained after 30 years; 60 years later the number roughly halved again, to 250 atoms, and after 90 years the number of atoms left was 108. This atom therefore has a half-life of 30 years. The curve shows the expected number of atoms remaining, based upon the idealized case in which exactly half the atoms decay in 30 years (it is, in other words, an exponential decay).

The randomness of radioactive decay is curious. For example, we might suppose that a freshly created nucleus would tend to last longer than an older one. But this is not the case – the decay of a nucleus is totally random and without any dependence on how the nucleus was created or its history. We now understand that this randomness is a fundamental feature of the Universe: it is a defining characteristic of quantum physics.

Like alpha decay, beta decay is also exploited in everyday life, for instance in PET scanners, where anti-matter is exploited to image the human body. Fluorine-18 is unstable to beta decay, which means that it is liable to convert into oxygen-18 with a half-life of just under 2 hours. In this case, the decay process involves the conversion of a proton inside the fluorine nucleus into a neutron, with the concurrent emission of an anticlectron (also known as a positron) and an electron neutrino. Positrons are identical to electrons, with the sole exception that they have positive electric charge. Crucially for PET scanning, when the

---

[1] Neutrinos are produced in vast numbers in the Sun, but most of them pass through ordinary matter as if it does not exist. Since they are so elusive, it took until 1956 before they were finally detected. Today there are several neutrino laboratories around the world.

BOX 3. EVERYTHING IS MADE OF ATOMS   P. 47

positron bumps into an electron the two annihilate each other with the production of two photons (particles of light). The photons have a lot more momentum than the original electron and positron, and so they travel away from each other in opposite directions. By surrounding the patient with a photon detector, it is possible to detect the individual photons and ascertain whereabouts in the body they were produced. Fluorine-18 is particularly useful for mapping out brain function or locating glucose-hungry cancer cells, because it can be incorporated into glucose molecules. Other positron emitters can be incorporated into a variety of molecules to trace out different biologically active regions of the body. The way these radioactive substances, such as fluorine-18, are produced is an interesting example of how blue-skies research often becomes relevant to everyday life. Specifically, fluorine-18 is manufactured using room-sized particle physics accelerators, by bombarding oxygen-18 with protons that have been accelerated through a voltage of a few million volts. PET scanners are also a good illustration of Einstein's $E = mc^2$ in action, because the mass associated with the initial electron and positron is entirely converted into the energy of the photons: each and every photon that is detected has energy equal to the mass of an electron multiplied by the speed of light squared (511 keV). This energy is large enough to guarantee that the two photons will travel away from each other in opposite directions. This, combined with the fact that all of the PET photons carry the same energy, helps in the detection process. PET scanners are beautiful examples of highly esoteric fundamental physics in everyday life.

The periodic table orders the atoms according to the number of protons in their nucleus, which is equal to the number of electrons surrounding it: this is called the 'atomic number'. The mass of each atom is also listed in the table, described in units where the mass of a proton and a neutron are approximately equal to 1. All of which means we can usually figure out how many neutrons are in an atom. The number of neutrons, which should be an integer, should be approximately equal to the atomic mass minus the number of

BOX 3. EVERYTHING IS MADE OF ATOMS   P. 48

protons. Gold has an atomic mass of 196.97 and contains 79 protons: an atom of gold should by implication contain 118 neutrons. There are some peculiarities, though. Take a look at chlorine (atomic number 17). It has an atomic mass of 35.453, which is midway between 35 and 36, so it seems that chlorine should contain around 18.5 neutrons, which does not make sense because there can only be an integer number of neutrons. The reason for this oddity is that chlorine atoms come mostly in two types: one type contains 18 neutrons and the other contains 20 neutrons. The lighter variation accounts for 76% of the mass in naturally occurring chlorine; the heavier one accounts for the remaining 24%. The mass quoted in the periodic table is the average of these two, i.e. $76\% \times 35 + 24\% \times 37 = 35.5$.

We have already been making reference to the atomic mass number, for example americium-241 is built from a total of 241 protons and neutrons. Various types of the same atom with different numbers of neutrons are called isotopes. We say that naturally occurring chlorine is composed mainly from two isotopes: chlorine-35 (often written ^{35}Cl) and chlorine-37 (^{37}Cl). As far as the chemistry is concerned, isotopes of the same element behave identically, since the chemistry only cares about the electrons. In contrast, the nuclear properties of different isotopes can be very different: fluorine-18 is a positron emitter, while fluorine-19 nuclei are stable, and therefore useless in PET scanners. The periodic table is of primary interest to chemists, which is why the information on isotopes is not explicit.

BOX 4. HOW BIG IS AN ATOM? P. 49

We can estimate the size of an atom by carefully dropping a tiny amount of oil onto the surface of some water and then measuring how much it spreads out. The oil will spread out because there is nothing to stop it from spreading, and it will form a layer a few atoms thick. This idea is attributed to the prolific physicist Lord Rayleigh who, on being awarded the 1902 Order of Merit at the coronation of King Edward VII, said: 'The only merit of which I personally am conscious was that of having pleased myself by my studies, and any results that may be due to my researches were owing to the fact that it has been a pleasure for me to become a physicist.' Without knowing more about how oil is made from atoms we do not know how many atoms thick the layer is, but we certainly know that the layer cannot be smaller than one atom thick. This means that by determining the thickness of the layer we can make a statement about the largest possible size of an atom.

So if we know how much oil we placed onto the water then we are in business. For example, a 0.5 mm diameter drop of olive oil spreads out over the water to a diameter of 25 cm. You can try this experiment at home and should measure the maximum extent of the oil drop, because water-softening agents might cause it to contract after a period of time as they attack the oil (so it would be better to use distilled water). The volume of olive oil is equal to $4\pi r^3/3$ where the radius $r = 0.25$ mm. This volume is also equal to the volume of the disk of oil on the water, which in turn is equal to $\pi R^2 d$, where $R = 12.5$ cm and d is the thickness of the disk. Equating these two volumes allows us to determine the thickness of the oil layer: $d = 1.3 \times 10^{-9}$ metres. So we know that an atom cannot be bigger than about 1 nanometre in size.

We can be a little more daring than this and make a stab at the size of an atom if we are prepared to accept that the layer of oil is one molecule thick and that each oil molecule is a chain of carbon and hydrogen atoms with one end attached to the water. These chains are typically around 10 atoms long (it depends on the type of oil). With this extra information, we can estimate that the size of one atom is roughly equal to 0.1 nanometres.

Knowing the size of an atom, we can ask how many are in a glass of water. If we assume that the atoms in a drop of

BOX 4. HOW BIG IS AN ATOM?   P. 50

olive oil are approximately the same distance apart as those in a drop of water, we can estimate the number of atoms in a glass full of water simply by dividing the volume of the glass by the volume of one atom. A 500 ml glass has a volume of half a litre, which we can divide by the volume of one atom to figure out how many atoms are in the water. 500 ml is 500 cubic centimetres, and we shall assume that each atom occupies a volume of approximately 1 cubic ångstrom (i.e. one atom fits inside a cube of side 1 ångstrom), which is 10^{-24} cubic centimetres. Taking the ratio informs us that there are something like $500 \times 10^{24} = 5 \times 10^{26}$ atoms in a half-litre glass of water. We can now use this information to figure out the mass of a water molecule. Our glass of water has a mass of 500 grams and we have estimated that it contains around 5×10^{26} atoms, so we can conclude that each atom weighs around 10^{-27} kilograms. All these numbers are not too far off the mark, and they are certainly within a factor of 10 of the true values – which is a terrific achievement given how many factors of 10 are involved.

There is a second way that we can estimate the size of an atom. We can use a more theoretical approach to figure out what we expect the answer to be. Let's focus on the simplest atom: that of hydrogen, with its one proton and one electron. The proton is very much heavier than the electron and can be thought of as providing an anchor about which the tethered electron dances. We want to know how far the electron is from the proton on average. The next paragraph is a little more mathematical than the norm for this book; skip it if you need to.

The distance that the electron is from the proton, d, can only depend on the size of the electrical charge of the proton and electron (Q), the electron mass (m) and a number that underpins the whole of the quantum world: the quantum of action ($\hbar$). The dependence on Q and m is fairly obvious: we expect that d should reduce as Q increases (the electron would be more tightly bound if it had more electric charge) or as m increases (the electron would be less inclined to fly away if it is heavy). The dependence on $\hbar$ is less obvious – but this is the one key parameter in quantum

BOX 4. HOW BIG IS AN ATOM?   P. 51

theory, and an electron dancing around a proton certainly is sensitive to quantum effects, so we should contemplate the fact that d might depend upon it. This is not the place for us to delve into the details of quantum mechanics; suffice to say that the quantum of action is what controls the wavelength of those electron waves in Figure 2.12. If we assume that these are the only quantities that d depends on, then we can say that d must be proportional to $\hbar^2/(mQ^2)$. We can be this confident because there is no other way to combine these quantities to give a quantity that can be measured in metres. The charge Q can be determined by ingeniously exploiting Coulomb's Law, which says that the force between two electric charges of size Q separated by a distance R is equal to Q^2/R^2. The result is that $Q = 4.8 \times 10^{-10}$ g$^{1/2}$ cm$^{3/2}$/s. The mass of an electron is $m = 9.11 \times 10^{-28}$ g and $\hbar = 1.1 \times 10^{-27}$ g cm^2/s. Putting these numbers in gives $d = 0.6$ ångstroms. We cannot claim an accuracy to better than a factor of a few using this method – but it should give us a rough answer. This way of figuring things out by appealing to the units carried by the salient quantities in a problem is used a lot by physicists, because it provides a quick way to estimate things that might be much more difficult to compute with precision. There is, of course, a proper (and much lengthier) way to do this calculation. As every undergraduate physics student knows, it involves solving the Schrödinger equation.

The number we just obtained is a factor of 6 larger than the number obtained by dropping oil on water. This really is very good agreement. So many powers of 10 are involved in the numbers we are working with that it would be very easy to get total disagreement if we did not understand things correctly. With no prior knowledge, except that provided directly by our senses, atoms could conceivably be any size at all smaller than something like a hundredth of a millimetre. That is a range spanning an infinite number of powers of 10, so the fact that the theoretical estimate and the oil-drop measurement give answers that agree to better than one power of 10 is very impressive.

3.
WEIGHING THE
EARTH

Our colleague Mike Seymour made an interesting observation while on holiday at Ogmore-by-Sea. Standing on
the mudflats by the water's edge, enjoying the cool of the
salty waves over his mildly sunburnt feet, Mike noticed a
buoy floating in the Bristol Channel that appeared to be
perched precisely on the horizon: an observation that is in
itself enough to make a rough estimate of the size of the
Earth. Being a physicist at rest,[1] Mike decided to gather
the necessary information. Releasing his heels with a
squelch he turned and walked with careful sharp-shell
cadence to a shop, and bought a map. This informed him
that the buoy, known as the Fairy Buoy, was approximately 4 km away from his vantage point on the beach,
which is marked by a red cross in Figure 3.1. A quick
sketch on the back of a seaside serviette (Figure 3.2)
allowed him to deduce that the Earth has a radius of
roughly 5000 km. The actual value is around 6400 km. It
may impress you that Mike made a reasonable estimate of
the size of our planet simply by observing the region
around Ogmore-by-Sea. Equally, you might be unim-
pressed that his answer is 20% out.

The calculation works on the assumption that the distance
to the buoy is in fact the distance to the horizon. The quality

---

[1] This is both grammatically ambiguous and, according to Einstein, physically ill-defined.

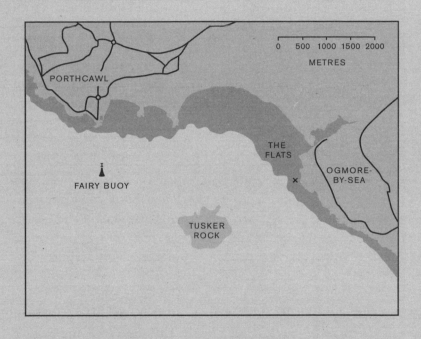

Figure 3.1 Mike stands on The Flats at Ogmore-by-Sea, at the position marked by a red cross. He estimates the distance to the horizon by noticing that the Fairy Buoy is approximately 4 km away.

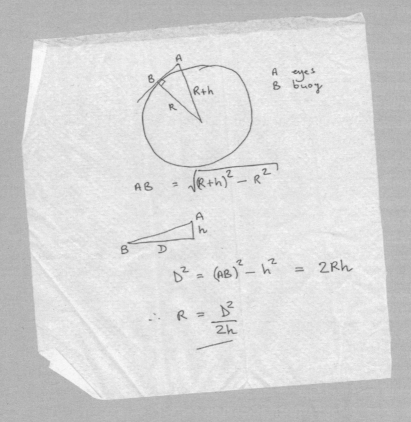

A eyes
B buoy

$$AB = \sqrt{(R+h)^2 - R^2}$$

$$D^2 = (AB)^2 - h^2 = 2Rh$$

$$\therefore \quad R = \frac{D^2}{2h}$$

Figure 3.2 Mike's eyes are $h = 1.6$ m above the surface of the Earth at point A. He estimates that the horizon vanishes at point B by observing the buoy perched perfectly in full view on the horizon. The map informs him that this is a distance $D = 4$ km from the beach. Using Pythagoras's Theorem, the radius of the Earth R is given by $D^2/(2h)$, where D is the distance between Mike and the buoy. Putting the numbers in gives $R = 5000$ km.

of Mike's eyesight governs how well he is able to determine whether or not the buoy is coincident with the horizon. A person with average eyesight can just about resolve a small coin at a distance of 40 metres, corresponding to an angular resolution of about 0.03 degrees. This means that Mike could perceive the buoy as being coincident with the horizon, even though the horizon is slightly in front of or slightly behind it. All he can say with certainty, therefore, is that the distance to the horizon lies somewhere between two extremes, defined by the resolution of his eyes, which he should quote as an uncertainty on the measurement. A little calculation[2] reveals that, given the limits of his eyesight, Mike really ought to have concluded that the radius of the Earth could quite easily be anywhere between 2000 km and 36,000 km. The fact that his serviette calculation got so close to the true value is largely a coincidence.

Estimating the uncertainty on a result is often as important as the result itself. It is only when we are aware of our ignorance that we can judge best how to use knowledge. In engineering or medical science, a deep understanding of uncertainty can be a matter of life and death. In politics, over-confidence is often the norm; uncertainty is seen as weakness when really it is a vital part of decision making. In this respect, science delivers an important lesson in humility.

In Mike's case, his measurement, while inaccurate, does

[2] If the base of the buoy was exactly on the horizon, then, using a little geometry, the horizon would be dipped below the horizontal level of Mike's eyes by an angle of 4 km/5000 km = 8×10^{-4} radians. Because Mike's eyes are not perfect, the actual dip to the horizon could be as small as 3×10^{-4} radians or as big as 13×10^{-4} radians. Now, a bit more geometry tells us that the radius of the Earth, R, is related to the dip angle by $R = 2h/(\text{dip angle})^2$. Putting in the two extreme dip angles gives R in the range 2000 km to 36,000 km.

still give us some idea about the size of the Earth. To achieve a better result, Mike would need to improve on the limiting resolution of his eyes, which can be done by using a camera with a long lens. Fortunately, Mike's dad, Bob, is a keen photographer, and lives in Ogmore-by-Sea. We couldn't make this up. We asked Bob if he might go down to the beach and take some photographs of the Fairy Buoy for us. He kindly obliged: a selection of his photos are shown in Figure 3.3.

Figure 3.3 Bob Seymour's photographs of the Fairy Buoy. They are all taken from the same place on the beach and with the same tripod height. The waves are causing the buoy to bob up and down.

Bob's photos were taken when the sea was quite choppy – which is perhaps a bonus, as we do not need to worry about the bending of light due to atmospheric effects, something which is more prone to happening on calm days: we can see that atmospheric effects are not an issue here because the images are pretty sharp. Bob adjusted the height of his camera such that the pictures show the buoy perched directly on the horizon. (Lowering the camera would push the buoy behind the horizon; raising it brings it in front of the horizon.) The photographs in the figure were taken at a height of 1.3 metres. The camera position was determined using GPS and the position of the Fairy Buoy was taken from the Trinity House[3] official records. The distance between buoy and camera was 4.15 km. These more refined numbers give a radius of the Earth equal to 6600 km. Using a camera has significantly reduced the uncertainty caused by limited resolution; the chief source of uncertainty that remains is now the difficulty in ascertaining the precise height of the camera above the average level of the waves. A 10 cm change in height leads to a 500 km change in the calculated radius of the Earth, which we might quote as a conservative estimate of the uncertainty on Bob's measurement.

There are, of course, far better ways of measuring the radius of the Earth than this – but that isn't the point. This is a good example of how simple, curiosity-driven observations, together with a little bit of careful thought, can lead to interesting conclusions. In what we suspect is a world first, Mike and his dad have measured the size of our planet from Ogmore-by-Sea. We have also learned a valuable lesson in quantifying uncertainty: it is easy to be misled into drawing the wrong conclusion unless we understand the degree of our ignorance.

In his measurement of the Earth, Mike followed in the footsteps of the greats (although if he'd actually stood on the shoulders of giants he wouldn't have been able to make his measurement). One of the earliest documented attempts to estimate the size of the Earth was made by Aristotle in 350 BCE. Aristotle noted in his book *On the Heavens* that 'there are stars seen in Egypt and in the neighbourhood of Cyprus which are not seen in the northerly regions,' and that the

---

[3] Trinity House, which dates back to the reign of Henry VIII, is responsible for the provision and maintenance of lighthouses and buoys in England, Wales, the Channel Islands and Gibraltar.

sphere of the Earth is therefore 'of no great size, for otherwise the effect of so slight a change of place would not be quickly apparent'. Using a very simple observation, Aristotle ruled out the possibility that the Earth has a radius much bigger than the distance between Egypt and the northern extent of the ancient world – that's to say, a few thousand kilometres. And, unsurprisingly for one of the most influential scientists ever, he was right. This is a terrific illustration of an 'order-of-magnitude' estimate. Order-of-magnitude estimates are quick calculations that are not supposed to be very accurate, and they are important in science because they can provide a good deal of insight with very little work.

The title of this chapter is 'Weighing the Earth', and that seems like a much taller order than estimating its radius. It is, but we can already make an order-of-magnitude estimate using the Seymour family measurements. Let's assume that the Earth is a perfect sphere. Its volume is $\frac{4}{3}\pi R^3$, which is 1.2 $\times 10^{21}$ m^3. In the absence of any other information, it isn't too crazy to suppose that the whole Earth might be a uniform sphere made of granite – a dense rock commonly occurring in the Earth's crust – or something whose average density is the same as that of granite. The density of granite is 2.8 grams/cubic centimetre, which gives us a very rough estimate for the mass of the Earth as 3.4×10^{24} kg. We have, of course, absolutely no way of knowing whether this is anywhere near right without finding some other way of weighing the Earth – and it is not at all obvious how this might be done. Let's work out how to do it.

We'll start by examining the motions of the planets across the night sky. At first glance this might seem to be a strange point of departure, but it will serve to illustrate an important point. Very often in science, work in one area can impact on superficially unrelated areas. This is one of many reasons why scientists should be allowed and encouraged to roam around researching anything that appears interesting. Nature is tremendously interconnected. Some time around 1510, the Polish astronomer Nicolaus Copernicus wrote a manuscript, the *Commentariolus*, in which he expressed his dissatisfaction with the classical Earth-centred cosmology of Ptolemy, formulated some 1400 years previously. 'I often consider,' Copernicus pondered, 'whether there could perhaps be found a more reasonable arrangement of circles, from which every apparent irregularity would be derived while everything in itself would move uniformly, as required by the rule of perfect motion.' The irregularities he was referring to are the occasional loops the planets perform, as viewed from Earth, as they make their way across the starry background over the course of weeks and months. Back in the second century CE, Ptolemy had devised a complicated system to predict the motion of the planets, which worked very well but was, at least to Copernicus's mind, ugly. In the *Commentariolus*, Copernicus asserts that the Moon goes around the Earth, the Earth and planets go around the Sun, that the daily motion of the Sun and stars is due to the rotation of the Earth on its axis, and that the distance from the Earth to the Sun is far smaller than the distances to the stars. He also suggests that the planetary loops we observe are a result of the Earth's motion relative to the planets. All of these statements are correct.

Copernicus published his complete works in 1543 in the six-volume *De Revolutionibus orbium coelestium* (*On the Revolutions of the Heavenly Spheres*), which is rightly regarded as one of the foundational early works in modern science. He showed that the complex motions of the planets can be understood if it is assumed that they all, Earth included, move in orbits around the Sun, each taking a different length of time to complete one circuit. Copernicus thought the orbits were circles but, as we now know, this is only an approximation. The planets follow slightly elliptical trajectories. The Earth takes 1 year to circle the Sun, Mercury takes 88 days, while Saturn makes the journey in 29 years. The Copernican Sun-centred model got a rough ride for many years, partly because the idea was seen to run counter to scripture by demoting the Earth from its previously imagined position at the centre of the Universe, and also because it doesn't feel as if the Earth is hurtling through space. This isn't a silly objection, and the deeper ramifications of the fact that we can't tell whether or not we are moving were only truly appreciated by Einstein in his special and general theories of relativity, published in 1905 and 1915. We will get to Einstein later.

If we accept Copernicus's wisdom, and arrange the planets in near circular orbits around the Sun, we can work out their distances to the Sun, in terms of the distance from the Earth to the Sun. For the inner planets, Mercury and Venus, the orbital radius can be calculated from a measurement of the largest angular separation between the planet and the Sun. (You can see this in Figure 3.4.) We defer to the next chapter the task of measuring the distance to Neptune, one of the outer planets, which we do explicitly using a digital camera, a good tripod and some photo-editing software. The orbital periods of the planets – the time it takes them to orbit once around the Sun – are also easily determined. In Figure 3.5 we show how to determine the length of time it takes for Jupiter to orbit the Sun by measuring the time between successive occasions when the Sun, Earth and the planet line up in the sky.

Just over half a century after Copernicus, the German astronomer Johannes Kepler spotted that there is a simple relationship between the size of a planet's orbit and the time taken to travel once around the Sun. Using data collected by Tycho Brahe, a Danish nobleman and fellow astronomer, Kepler noticed that the square of the orbital period (or T^2 for short) is proportional to the cube of the radius of the orbit (R^3 – more precisely, it is the cube of the semi-major axis of an elliptical orbit). This means that the ratio of T^2/R^3 should be the same for each planet and, as the data in Table 3.1 illustrate, Kepler was on to something.

PLANET	T (years)	R (Earth radii)	T^2/R^3
Mercury	0.241	0.387	1.00
Venus	0.615	0.723	1.00
Earth	1	1	1
Mars	1.881	1.524	1.00
Jupiter	11.86	5.203	1.00
Saturn	29.46	9.555	0.99
Uranus	84.02	19.22	0.99
Neptune	164.8	30.11	0.99

Table 3.1 The orbital parameters of the planets in the solar system.

In 1687, the first edition of Isaac Newton's *Philosophiæ Naturalis Principia Mathematica* was published. In it, Newton demonstrated that the empirical pattern Kepler had detected is a consequence of a deeper physical law. The

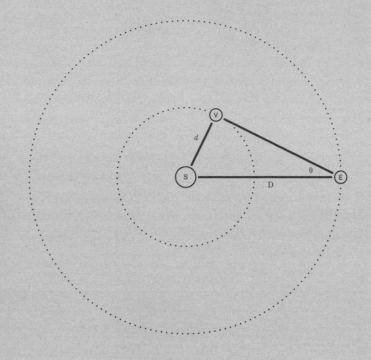

Figure 3.4 We can determine the distance from Earth to Venus or Mercury by measuring the angle, θ, which is the largest angular distance in the sky between the planet and the Sun. The distance d is then obtained by basic trigonometry, i.e. $d = D \sin\theta$. These methods allow all distances to be computed relative to the distance between the Earth and the Sun. The distance between Earth and Venus can also be measured by timing how long it takes for a radar signal to travel to Venus and back. Once we know that distance, we can determine the orbital distances of all of the planets in the solar system in metres. This radar measurement was first achieved in 1961, using the twin 26-metre-diameter radio telescopes at the Goldstone observatory in the Mojave Desert, California. The result gave a value for the mean Earth–Sun distance of 149,599,000 km. It's quite hard to base a high-precision measurement system on something defined simply as 'the mean distance from the Earth to the Sun'. So, in August 2012, the International Astronomical Union defined the Astronomical Unit (AU) to be precisely 149,597,870,700 metres. This means that the mean distance from the Earth to the Sun is not precisely 1 AU any more, but it never was precisely anything anyway. The precision of modern measurement has meant that astronomers have outgrown the old, intuitive definition.

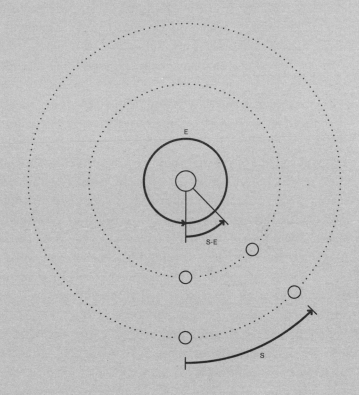

Figure 3.5 Measuring the time it takes for an outer planet to orbit the Sun. S is the time between successive occasions when the Sun, Earth and the planet all line up (astronomers say that the planets are 'in opposition' when they line up like this). Simple geometry gives (S-E)/E = S/P, where P is the time it takes for the planet to orbit the Sun and E is the time it takes the Earth to orbit the Sun, i.e. 1 year. This implies that $1/E - 1/S = 1/P$, a formula Copernicus used five centuries ago in *De Revolutionibus*.

idea that regularities and patterns in Nature are often the sign of an underlying simplicity that can be captured by mathematical equations is a familiar one to scientists today – but in the late seventeenth century, Newton's discovery was revolutionary.

The mathematical, law-based approach that Newton introduced in his *Principia* is the foundation for virtually all of modern physics. He showed that Kepler's T^2/R^3 pattern is a consequence of the existence of a Universal Law of Gravitation, which states that all massive bodies attract each other with a force that is proportional to the product of their masses, and inversely proportional to the square of the distance between them. For a planet orbiting the Sun, the law states that the Sun exerts a force F on the planet, and that $F = GMm/R^2$, where M and m are the masses of the Sun and the planet respectively, and R is the distance between their centres. The quantity G is now known as the Gravitational Constant, and it is a number whose value describes the strength of the gravitational force. Newton also introduced what is now referred to as his Second Law of Motion, which describes how an object – such as a planet – moves when a force acts upon it. This Second Law of Motion states that forces induce accelerations according to the equation $F = ma$, where m, in our case, would be the mass of the planet and a is the acceleration. With just these two equations, it is possible to understand the origin of the elliptical planetary orbits and Kepler's T^2/R^3 pattern. Box 5 shows details for the simplest case of a circular orbit.

It is hard to overstate the radical leap forward delivered by Newton, and we could spend the rest of this book exploring the consequences of his laws. But in this chapter we're

BOX 5. KEPLER'S LAW   P. 65

Newton's law of gravity, together with his equation $F = ma$, explain Kepler's law. This is easy to understand in a case where the planet's orbit is a perfect circle but, with a little more maths, it also works for elliptical orbits. Here we show how Newton's law works out for a circular orbit. Since $F = ma = GMm/R^2$, it immediately follows that the planet accelerates towards the Sun with an acceleration of GM/R^2. But for things that go in circles at constant speed, this acceleration must also be equal to v^2/R, where v is the speed of the planet. Equating these two accelerations gives $GM/R = v^2$. But v is related to the time T it takes for the planet to orbit the Sun by $v = 2\pi R/T$, which means that $GM/R = 4\pi^2 R^2/T^2$. This tells us that $T^2/R^3 = 4\pi^2/(GM)$, which is constant. In fact, what we have done is only approximately correct. The mass M appearing in Kepler's law really should be $M+m$. This is because the force F also acts on the Sun, causing it to accelerate too. This is a small effect if M is much bigger than m, which is the case for all of the planets in the solar system. Generally speaking, two objects will orbit about a point a fraction $m/(M+m)$ along the line joining them. For example, two equal mass objects will orbit around a point midway between the two.

focused on one specific goal. We want to weigh the Earth[4] –
and in this context, Newton's laws offer something extremely
important. They relate the motion of something, such as the
orbit of a moon around a planet or a planet around the Sun,
to the mass of the thing that induces that motion, through
the Universal Law of Gravitation. In the language of a physicist, this is highly non-trivial.

Before Newton, there was no known connection between
these things. All of Newton's laws are *universal*, which means
that they don't only apply to moons and planets and stars.
They are supposed to apply to *any* objects, *anywhere*. This is
also a highly non-trivial statement, because it means there is
a common framework that can describe the motion of the
planets in the heavens and the motion of objects like cannon-
balls and swinging pendulums here on Earth. Perhaps you
can see where we are heading.

Let's consider a ball falling to the ground. How is this
described using Newton's laws? The force acting on the ball,
accelerating it towards the ground, is given by $F = GMm/r^2$,
where M is the mass of the Earth, m is the mass of the ball,
and r is the distance between the centre of the Earth and
the centre of the ball. The way the ball responds to this force
is given by Newton's Second Law of Motion, $F = ma$. A
very simple bit of algebra gives us the acceleration of the ball
induced by the gravitational pull of the Earth: $a = GM/r^2$.
Using a ruler and a watch, we can measure the acceleration
of a ball when we drop it (we'll get close to 9.8 m/s^2) and,

---

[4] Strictly speaking, we want to determine the mass of the Earth. Weight is a measure of
how much a mass is pulled under gravity. Your weight would be different on Earth or
on the Moon, even though your mass is the same.

with $r = 6370$ km, we can calculate the product of the mass and the Gravitational Constant: $GM = 4.0 \times 10^{14}$ m^3/s^2.

Or consider the orbit of the Moon around the Earth. We can measure the average distance from the Earth to the Moon,[5] R (385,000 km), and the period of the Moon's orbit, T (27.3 days). Newton's Universal Law of Gravitation relates these quantities to the mass of the Earth, M, through the equation we derived in Box 5: $T^2/R^3 = 4\pi^2/(GM)$. As in the case of a dropped ball, we have a way of determining the product GM, only this time from astronomical observations. Putting the numbers in gives $GM = 4.0 \times 10^{14}$ m^3/s^2, as before. We get the same answer for a dropped ball and for the orbiting Moon because Newton's laws are universal; they encode information about the deeper physical structure of our universe.

Armed with the value of GM, we can go ahead and determine either the mass of the Earth (M) or the gravitational constant (G), but only if we already know the value of one of them. So we need some new method to measure G or M, but this is not an easy thing to do, because gravity is a colossally weak force. In theory, a simple way to determine the mass of the Earth would be to measure the amount by which a hanging plumb-line is attracted towards a large mass placed next to it, as in Figure 3.6. If we position a ball of lead with a mass of 1000 tonnes and a radius just under 3 metres so that its centre is horizontally aligned with a hanging mass placed 3 metres away, the plumb-line will be attracted to the ball by a minus-

---

[5] Historically the distance to the Moon was obtained by the parallax method described in the following chapter, but today it's done to sub-millimetre precision by bouncing laser light off mirrors left on the Moon by the Apollo astronauts.

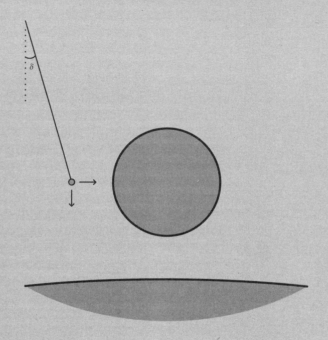

Figure 3.6 A hanging plumb-line next to a large sphere. The tangent of the angle δ is the ratio of the horizontal pull of the ball to the vertical pull of the Earth. According to Newton's Universal Law of Gravitation, this is just $(Gm/r^2)/(GM/R^2)$ where R is the radius of the Earth. The dependence on G cancels, giving $\tan \delta = (m/M) \times (R/r)^2$.

cule angle of 0.16 arcseconds from the vertical[6] (1 arcsecond is an angle equal to 1/3600th of a degree). But although the idea is theoretically simple, it is clearly not an easy thing to do in practice. It isn't cheap, either: a lead ball this big would cost around a million pounds in today's money.

Two classic experiments, performed by Nevil Maskelyne and Henry Cavendish in 1774 and 1798 respectively, were the first to allow for an accurate determination of G. It is testament to the difficulty of the measurements that a century passed after Newton published his theory of gravity before anyone managed to perform experiments capable of determining its strength. We've been careful with our wording here because, for historical reasons, neither Maskelyne nor Cavendish actually quoted the value of G; they were both only interested in weighing the Earth. From a physicist's perspective this doesn't matter at all, because once you have one quantity you can get the other in a single line of mathematics.

In 1774, the Reverend Nevil Maskelyne, Astronomer Royal, led a team to Perthshire, in the Scottish Highlands, to measure the deflection of a plumb-line in the vicinity of a mountain called Schiehallion. There, he carried out what is in essence the experiment illustrated in Figure 3.6 – except that he replaced the 1000-tonne ball by an entire mountain. This magnified the deflection of the plumb-line, making it (just about) measurable. Using the stars as reference points, he succeeded in measuring a very small deflection of 11.6 arcseconds (in fact, this was the sum of two deflections

---

[6] This angle can be obtained using the formula in the figure caption if we anticipate that the Earth's mass is 6×10^{24} kg. Of course the logic is the other way around, i.e. we are proposing to use the measured deflection to infer the Earth's mass.

corresponding to the plumb-line being located on the north and south sides of the mountain). Four years later, following a detailed survey of the mountain, mathematician Charles Hutton used Maskelyne's measured deflection to estimate the density of the Earth. It was, he calculated, 4.5 times that of water, which is quite a bit higher than the density of rock. Assuming that the Earth is a perfect sphere, and using the modern measurement of its average radius, 6370 km, this gives us a mass of the Earth equal to 4.9×10^{24} kg.

The next major step in weighing the Earth was taken some twenty years later by the brilliant Henry Cavendish, using a remarkable experiment in his very large house overlooking Clapham Common, in south London.

Cavendish was an eccentric man of independent means; he wore old-fashioned clothes and was famously very shy. And he was one of the greatest scientists of all time – apart from the work we describe here, he made hugely important contributions in the fields of chemistry and electricity. Today, Cambridge University's Department of Physics, the Cavendish Laboratory, is named after him. In 1798, Cavendish detailed the results of his experiment – including a drawing of the apparatus he devised for it – in his 'Experiments to Determine the Density of the Earth'. He begins by crediting the ex-Cambridge professor of geology Reverend John Michell for designing the experiment and building the first version of it; Michell, though, died in 1793, before he could make any measurements, and the apparatus found its way to Cavendish via another Cambridge philosopher-priest, the Reverend Francis John Hyde Wollaston.

The idea behind the experiment is very simple, although the precision with which Cavendish made his measurements required the touch of a brilliant experimenter. Two small balls are hung from either end of a long beam, which is itself suspended by a thin wire. Two much bigger balls are then moved close to the small balls, one on each side of the beam. According to Newton's theory, the large balls will exert a gravitational force on the small balls, thereby causing the beam to twist slightly. Cavendish measured the twist, and used the measurement to determine the force that the large balls exert on the small ones.[7] To avoid any external disturbances, he put the apparatus in a closed room. The 6-foot wooden beam was suspended from the point marked F in Figure 3.7, and the two small lead balls of 2-inch diameter were hung by threads inside the boxes marked DCB. Cavendish was able to measure the twist of the beam using two telescopes T trained on ivory scales marked in gradations of hundredths of an inch. The larger 12-inch diameter lead balls, labelled W, were hung by copper rods (labelled r at the top and R at the bottom) and moved into place by a pulley system operated from outside the room. By making it possible to operate the apparatus remotely, Cavendish improved Michell's original design substantially, isolating his apparatus as much as possible from unwanted disturbances.

We've gone into some detail because we love this sort of thing. Cavendish's experiment is very modern in many ways. It is a high-precision apparatus designed to measure the tiny gravitational force exerted by the heavy balls – a force that amounts to no more than the weight of a grain

---

[7] The force can be deduced by measuring the wire's resistance to being twisted. This is done by timing the back-and-forth oscillations of the beam that occur when it is twisted without the presence of the heavy balls.

Figure 3.7 The layout of Henry Cavendish's
ingenious apparatus to weigh the Earth.

of sand. Success depended on Cavendish exercising extreme care, observing with extreme accuracy, and developing a solid understanding of the principal sources of uncertainty. He considered the effects of magnetism, temperature variations in the room, variability in the stiffness of the suspension wire, the gravitational pull of the rods from which the heavy balls were suspended, air currents, the gravitational attraction of the wooden case on the balls and the beams, and more. He operated according to the same scientific principles as the Seymours at Ogmore-by-Sea, but with a quite stunning attention to detail that was mandated by the delicacy and subtlety of the property of Nature he wished to measure.

In the final analysis, he determined the mean density of the Earth to be 5.45 times that of water, i.e. 5.45 grams/cm^3. As Cavendish noted, in respectfully circumspect tones:

> According to the experiments made by Dr. Maskelyne, on the attraction of the hill Schehallien [sic], the density of the earth is 4½ times that of water; which differs rather more from [my measurement] than I should have expected. But I forbear entering into any consideration of which determination is most to be depended on, till I have examined more carefully how much [my] determination is affected by irregularities whose quantity I cannot measure.

Cavendish was taking care not to rush to conclusions but, as future measurements revealed, he was bang on the money.

Charles Hutton, the mathematician who had weighed the Earth using the measurements from the Schiehallion experiment, remained extremely sceptical of Cavendish's more delicate measurement to the day he died. In 1801, encouraged by Hutton, the Scottish philosopher-mathematician John Playfair conducted a new lithological survey of Schiehallion and, in 1811, revised Hutton's original measurement upwards to 4.71. Two years before his death in 1823, and 11 years after Cavendish had died, Hutton made his views very clear in his paper 'On the Mean Density of the Earth'. 'From the closest and most scrupulous attention,' he wrote, 'the preference, in point of accuracy, appears to be decidedly in favour of the large mountain experiment over that of the small balls.' He spoke condescendingly of Cavendish's 'pretty and amusing little experiment' and sniffed at the veracity of results 'produced by machinery so complex' and 'calculated by theorems derived from intricate mathematical investigations'.

Cavendish, though, was right. Today, the best measurement of the mean specific density of the Earth is 5.515, which is only 1.2% different from Cavendish's result. Cavendish's balls may have been small, but he more than made up for it with careful attention to detail and a steady hand.

There is an interesting postscript to this story. In 2007, an expert in oil and gas exploration called John Smallwood revisited the Schiehallion measurement, using modern methods to determine the geometry and composition of the mountain. He was provoked by a challenge issued in Charles Hutton's 1821 paper, in which Hutton stated categorically that the Earth was 'very near five times the density of water; but not higher', before throwing down the gauntlet: 'Let any person, who doubts, look over and repeat the calculations ... and try if he can to find an inaccuracy in them.' Smallwood did just that. His re-analysis, published in the *Scottish Journal of Geology*, concluded that Maskelyne's original measurement allows one to conclude that the specific density of the Earth is 5.48 with an uncertainty of 0.25.

We presented Cavendish's number as he published it, in terms of the average density of the Earth. His measurement corresponds to a mass for the Earth of 5.90×10^{24} kg. The modern value is 5.97×10^{24} kg. What a ringing endorsement of the brilliance of Cavendish. If you recall, the Seymour family measurement, together with our guess that the Earth is a uniform sphere of granite, gave us a mass of 3.4×10^{24} kg. Not bad for a day on the beach.

Actually, things get even better than 'merely' measuring the mass of the Earth. Because now that we have the mass of the Earth we can determine the gravitational constant, $G = 4.0 \times 10^{14}$ m^3/s^2/5.97×10^{24} kg $= 6.7 \times 10^{-11}$ m^3/s^2/kg. And with G we hit the jackpot, because it means that now we can weigh anything in the Universe that has something orbiting around it. For example, since we also know that the Earth orbits the Sun with a period $T = 365.25$ days at a distance $R = 150$ million km, we can deduce (using the formula in Box 5) that the Sun is about 330,000 times more massive than the Earth. And we can keep going.

Jupiter has a large collection of moons, the brightest of which were discovered by Galileo in 1610. By observing their orbits, we have measured the mass of Jupiter to be 1.898×10^{27} kg; a colossal gas giant world 318 times more massive than the Earth. We can observe clouds of gas, glowing in the radio spectrum, orbiting around distant galaxies, and this allows us to measure the mass of the galaxies. Andromeda, the nearest galaxy to the Milky Way, is 1.5 trillion times the mass of the Sun. Only a small fraction of this mass is visible to us in the form of stars, however, which has led astronomers to suggest that there is an ocean of unseen dark matter, probably in the form of new, as yet undiscovered subatomic particles, permeating the galaxies. There is now a great deal of independent evidence for the existence of dark matter from studies of the evolution of the Universe and of Cosmic Microwave Background radiation (we'll look at this in much more detail later on), but we should point out here that dark matter was first discovered using Newton's laws and our knowledge of G. And, perhaps strangest of all, at the heart of the Milky Way galaxy in the direction of the constellation of Sagittarius, there are stars known as the S-stars with extreme orbits around a dense, compact object. Using Newton's laws, the object is measured to be over 4 million times the mass of our Sun. Astronomers believe this exotic object to be a supermassive black hole, 26,000 light years from Earth, devouring dust and gas from the rich clouds that drift through the dense central regions and spewing radiation out across the galaxy.

Newton's laws are treasures. They were the first universal laws of Nature to be discovered, and they have allowed us to discover unimaginable things unfathomably far away, starting from the beach at Ogmore-by-Sea.

4.
THE DISTANCE TO
THE STARS

The stars are tiny specks of light in the darkness. Unlike the planets, they exhibit no extravagant loops on the sky, and, for the overwhelming majority of them, we have no telescope capable of resolving any detail on their surfaces. Beyond the nightly circular arcs across the sky induced by the Earth's spin, they appear to be immobile, featureless points. And yet we know the distance to each and every one. As we will see later in the book, this ability to map the cosmos with precision is the Rosetta Stone that will allow us to explore the Universe's origin and evolution: the skies are teeming with information and we have learned how to decode it.

Before we cast our gaze beyond the solar system, we'd like to begin by making good on our promise in the last chapter to measure the distance to the most distant planet in our solar system: Neptune. The method we'll use is also the method by which we determine the distances to the nearest stars. It is called the parallax method.

Parallax is an effect with which everyone is familiar; it is the reason humans have two eyes. If you hold one of your fingers up in front of your face, and alternately close one eye and then the other, your finger will appear to shift against the background. The closer your finger is to your face, the greater the shift. As I (Jeff) am writing this I have decided to use parallax to measure the length of my own arm. I (Brian) am not in the least surprised. With my arm extended, I look at my index finger, first with my left eye closed and then with my right eye closed. I notice that my finger moves by about 8 degrees. I've downloaded the image of a large protractor onto my laptop for the job. Figure 4.1 shows the geometry, and, knowing that the distance between my eyes – which I measured with a ruler – is about 6.5 cm, I have deduced that my finger was located a distance 3.25/tan(4°) = 46 cm away. The parallax shift of nearby objects, discernible because we have two separated eyes, is one of the pieces of information our brains exploit to estimate distances. It is a two-eyed ability that has been selected for in many animals.

Unfortunately, you can't estimate the distance to Neptune by winking: both because your head isn't big enough, and because Neptune is not visible to the naked eye. In fact, it is so faint that it wasn't discovered until 1846, when the French mathematician Urbain Le Verrier made a prediction of a new planet's position in the sky using Newton's laws, based on his own observations of irregularities in the orbit of Uranus. He sent the prediction to the German astronomer Johan Gottfried Galle, who on the evening of 23 September duly found it after only an hour of searching. Fortunately, Le Verrier was able to circumvent his lack of a sufficiently large head – and so too can we, as Figure 4.2 illustrates.

The position of Neptune against the fixed background stars changes due to parallax as the Earth moves around the Sun, providing us with different vantage points every evening. This figure shows the situation when Neptune is 'in opposition', which means it is lined up with the Earth and the Sun. We chose to make the measurement close to opposition because it makes the mathematics easier, but it is not hard to perform the calculation at any other time. In 2014, this planetary alignment occurred on Friday 29 August, and we asked Kevin Kilburn, an amateur astronomer with the Manchester Astronomical Society, to photograph Neptune for us around the time of opposition. He obtained four photographs on 19 August, 29 August, 9 September and 21 September – very probably the only clear nights in late-summer Manchester that year. We superimposed the photographs on top of each other, lining them up using the background stars; the resulting image is shown in Figure 4.3 (p. 87).

The procedure for determining the distance to Neptune is precisely the same as the one Jeff used to calculate the length of his arm. We need to know the 'distance between the eyes', which is the distance the Earth moves from one photograph to the next, and the angle through which Neptune shifts against the starry background during this time, which we can get from the photographs. We describe every step of this process in detail in Box 6 (pp. 86–90), and we hope you decide to follow it through, because it's a lovely, very simple measurement that you might choose to perform for yourself. Our calculation, using only Kevin's photographs, gives the distance of Neptune from the Sun as 30.33 Astronomical Units (AU). The official figures tell us that Neptune was at a distance of 29.96 AU from the Sun on 29 August 2014, which differs from our measurement by about 1%.

This parallax method can only deliver the distance to Neptune in terms of the Earth–Sun distance (which is 1 AU). If we want to express astronomical distances in terms of metres then we need to figure out what 1 AU actually is in metres. In the last chapter we dealt with this by noting that we can ascertain the radius of Venus' orbit around the Sun by two different methods. First (as illustrated in Figure 3.4)

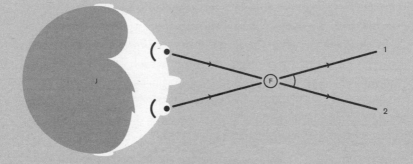

Figure 4.1 Overhead view of
Jeff's head and finger. Direction
1 corresponds to the direction
in which his finger appears when
his left eye is closed, direction 2
when his right eye is closed.

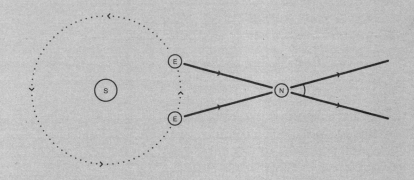

Figure 4.2 Just like the previous figure, except the eyes are replaced by two positions of the Earth in its orbit around the Sun and the finger is replaced by Neptune. By measuring how much Neptune shifts its angular position with respect to the distant stars (they are so far away that their angular deflection is much smaller than Neptune's), we can deduce the distance to Neptune.

using trigonometry (the result is around 0.7 AU) and, second, by bouncing radar off the planet (the result is 108 million km). That information is sufficient to establish that 1 AU is around 150 million km. The Astronomical Unit is one of the most important numbers in astronomy, because it unlocks the whole distance ladder. What's more, the pre-radar quest for a precision measurement of an absolute distance between any two celestial objects is one of the great stories in the history of science, spanning many centuries.

In determining distances using parallax, we use two vantage points on the Earth's surface, or two points on the Earth's orbit around the Sun – the equivalent of two eyes. For example, on a clear night, we could take two photographs at precisely the same time from two different points on the surface of the Earth. If we know the distance between the cameras, then we could use parallax to determine the distance to the object we photographed. This would be difficult to do for Neptune, though, because it is too far away and the parallax shift would be very small. For this reason, in our measurement of the distance to Neptune, we increased the distance between the two vantage points by waiting for the Earth to move around the Sun. But measuring distances using the view from two different points on Earth is viable for less distant planets.

In 1672, when Mars was in opposition, the great Italian astronomer Giovanni Cassini and his colleague Jean Richer made a measurement for the parallax of Mars. Richer observed the position of Mars from French Guiana in South America, and Cassini worked from Paris. They knew the 'distance between the eyes' with reasonable precision, and from this they were able to calculate that the Astronomical

Unit is approximately 21,700 Earth radii. It was a pretty good attempt. The modern value is 23,455 Earth radii.

There is a whole book to be written on the numerous expeditions the explorers and astronomers of the seventeenth, eighteenth and nineteenth centuries embarked upon to refine the measurement of the Astronomical Unit. We aren't writing that book, but we cannot resist mentioning the story of the French astronomer Guillaume Le Gentil. One of the classic, high-precision parallax measurements can be made during a transit of Venus, which is when the planet crosses the face of the Sun as seen from Earth. Because of parallax, observations of the transit from two different points on the Earth will differ, something which can be exploited to ascertain the distance to Venus. Transits of Venus come in pairs, 8 years apart, and then do not repeat for over a century. The most recent transit occurred in June 2012; there will not be another one until December 2117. So if you really want to make a precision measurement of the Astronomical Unit, you don't want to miss it – and, back in the mid-eighteenth century, Guillaume Le Gentil certainly didn't.

In March 1760, Le Gentil left Paris and headed for the Indian city of Pondicherry as one of an international team of over a hundred observers sent out across the planet to make multiple observations of the Venus transit of 1761. He made it to Mauritius, then known as the Isle de France, in July of that year, but one of the regular wars between France and Britain made his onward journey too dangerous. Finally, in March 1761 he managed to board a ship bound for India, and although the ship was blown off course he still made it to Pondicherry with days to spare. Unfortunately, the British had occupied Pondicherry and the ship couldn't land, so the captain swung it around and headed back to Mauritius. The day of the transit was beautifully clear, but with the ship still in open sea, pitching and rolling, precise astronomical observations were impossible.

Undaunted, Le Gentil decided to wait eight years in and around the Indian Ocean for the transit of 1769, and after spending some time mapping the coast of Madagascar he headed for Manila in the Philippines. But the Spanish in Manila were not helpful, so he returned to Pondicherry and built an observatory in readiness for his vital moment. When the morning of 4 June 1769 duly arrived, it turned out to be the only cloudy day in weeks, and he missed the transit. After a few miserable months, he finally boarded a ship bound for home, but an outbreak of dysentery and severe storms resulted in him being dropped off at La Réunion, off the eastern coast of Madagascar. He was finally able to board a Spanish ship to take him back to Paris, where he arrived in October 1771 to discover that he had been declared legally dead, his wife had remarried, his relatives had sold his estate, and he had lost his position in the Royal Academy of Sciences, which had dispatched him on the expedition in the first place. This must surely have been the moment for which the phrase 'For fuck's sake!' was invented. To put your mind at rest, Le Gentil was reinstated at the Royal Academy of Sciences, remarried and lived a happy life for a further twenty-one years.

This story goes to show the sheer commitment of these pioneers, who understood the importance of measuring the distance between the Earth and the Sun and thereby unlocking the distance scale that allows us to measure the Universe today. These scientists and explorers were brilliant, dedicated people, who – as the case of Le Gentil shows – spent their lives attempting to acquire the extremely hard-won knowledge upon which our understanding of the Universe now rests.

Using the parallax method, we made a precise measurement of the distance to Neptune, the most distant planet in the solar system. This raises the question: how far out into the cosmos can we go with the parallax method? Well, we can certainly measure the parallax shifts of the nearest stars. Because the Earth orbits the Sun, the closest stars will oscillate back and forth across a small angular region of the sky, and the extent of that back-and-forth motion allows us to determine how far the star is away. With this in mind, the maximum possible baseline[1] for a parallax measurement is the diameter of the Earth's orbit (2 AU), which corresponds to making measurements of the parallax shift of a star six months apart. The first determination of the distance to a star was made using this method in 1838 by the German mathematician and astronomer Friedrich Bessel, who observed that the star 61 Cygni has a parallax angle of 0.314 arcseconds (which means the extremes in its position in the sky are separated by 0.628 arcseconds, because astronomers define the parallax angle to be half the angular shift). Using this, he concluded that its distance from the Earth is 10.4 light years,[2] which is close to the modern value of 11.4 light years.

---

[1] This is the 'distance between the eyes', but we are getting a bit more professional with our language.

[2] If you have been following the details you can work this out for yourself: the tangent of 0.314 arcseconds is equal to the ratio of the Earth–Sun distance to the distance to the star. Since the angle is very small, the tangent is approximately equal to the angle expressed in radians. This means the distance is equal to 1 AU divided by $0.314/3600 \times \pi/180$.

BOX 6. MEASURING THE DISTANCE TO NEPTUNE   P. 86

Kevin Kilburn took the photographs of Neptune shown
in Figure 4.3 using a 300 mm Zeiss Pentacon lens on a
Canon 550D digital camera, attached to a Skywatcher EQ5
equatorial mount. The mount allows the camera to track
with the rotation of the Earth, meaning that a 20-second
exposure can be made without the image being blurred.
The four photographs in the figure were taken over the
course of a month in August and September 2014. We
used Adobe Photoshop to stack the photographs on top
of each other, such that all of the background stars line
up. The motion of Neptune across the sky is evident: it is
the only thing that moves. It is possible to make parallax
measurements of the brighter planets using virtually any
digital camera – and here, a normal fixed photographer's
tripod can be used, because the exposure times can be
much shorter. Despite needing slightly more expensive
gear, though, it's worth using Neptune; the maths is
slightly simpler, because Neptune's motion relative to the
background stars is mainly a result of the Earth's motion
around the Sun.

If we know the speed at which Neptune moves across
the sky when it is in opposition, then we can easily work out
how far it is away from the Sun. To see how this is done,
look again at Figure 4.2. For the moment we will ignore the
fact that Neptune is orbiting around the Sun. Neptune's
orbit won't affect our measurements too much, because
it takes 165 years to make one complete circle, so from
our perspective, won't move much over the course of a
month – but in any case we will correct for this later on.
The figure shows the Earth at two points in its orbit around
the Sun, marked A and B. These could represent two of
our photographs. The corresponding shift in the position
of Neptune against the stars is the angle marked a. The
distance between points A and B is approximately equal
to $(a/360°) \times 2\pi R$, where R is the distance from the Earth
to Neptune – which of course is what we want to know.
We can also calculate the distance between points A and
B another way: it is approximately equal to the distance
the Earth travels around the Sun in the time between

BOX 6. EVERYTHING IS MADE OF ATOMS   P. 87

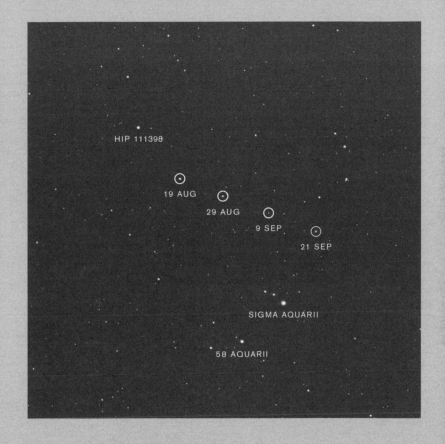

Figure 4.3 Neptune (circled) moving across the
sky relative to the stars in August/September
2014. Photographs by Kevin Kilburn of the
Manchester Astronomical Society.

BOX 6. MEASURING THE DISTANCE TO NEPTUNE   P. 88

photographs, i.e. (the circumference of the Earth's orbit around the Sun) × (the time between measurements) / (1 year). This means that

$R =$ (the time between measurements) / (1 year) / (the angle a by which Neptune moved across the sky between measurements / 360 degrees) × 1 AU

All that remains is to determine a from the photographs. The table below lists the pixel positions of Neptune, obtained by loading the 5184 × 3456 pixel images into Photoshop. We will focus on the first two measurements, taken on 19 and 29 August, the latter being the day when Neptune was in opposition. We need to convert the change in pixel co-ordinates into the angle through which Neptune moved in the sky. The change in the horizontal pixel co-ordinate is $(2950 - 2640) = 310$, and the change in the vertical pixel co-ordinate is $(1756 - 1629) = 127$. We can convert this into an angle if we know the image's field of view, by which we mean the angular portion of the sky covered by one of the photographs. We can determine the field of view as follows. The sensor on a Canon 550D camera is 22.3 mm × 14.9 mm across. If you decide to make your own measurement, you'll need to look in your camera's user manual to find the sensor size. Figure 4.4 shows two rays travelling from points in the sky into the 300 mm focal-length lens and landing on the very top and very bottom of the sensor.

The tangent of the angle marked θ (theta) is equal to (14.9 mm/2) / (300 mm), which means θ = 1.42°. This means

DATE	X (PIXELS)	Y (PIXELS)	TIME UT
19 AUGUST	2640	1629	2300
29 AUGUST	2950	1756	2348
9 SEPTEMBER	3285	1879	2303
21 SEPTEMBER	3638	2014	2110

Table 4.1 The four positions of Neptune as it moved across the sky during August and September 2014. The pixel co-ordinates were measured using Adobe Photoshop. The fact that they were taken at different times of day does not matter because the images were stacked on top of each other with the stars aligned.

BOX 6. EVERYTHING IS MADE OF ATOMS   P. 89

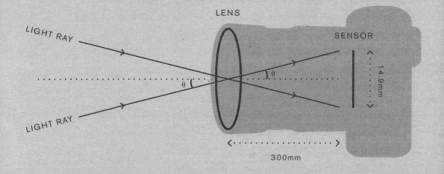

LENS

LIGHT RAY

SENSOR

14.9mm

θ

θ

LIGHT RAY

300mm

Figure 4.4 Determining the field of view of Kevin's camera.

BOX 6. MEASURING THE DISTANCE TO NEPTUNE   P. 90

that the top and bottom of any photograph taken with these settings has an angular spread of 2.845° degrees. We can do the same for light coming in from the left and right. In that case the tangent of the relevant angle is (22.3 mm / 2) / (300 mm), which gives an angular spread equal to 4.26°. Kevin's photos therefore cover a patch of the sky equal to 4.26 × 2.845 square degrees, which is approximately 1/3400th of the entire sky. The images shown in Figure 4.3 correspond to a zoomed-in portion of this. Because the change in angle is so small, we can use Pythagoras' Theorem to determine that, between those two dates, Neptune travelled across the sky through an angle equal to $\sqrt{((310/5184 \times 4.26)^2 + (127/3456 \times 2.845)^2)} = 0.2754$ degrees.

Armed with this angle we can use the underlined formula on the previous page to deduce that the distance from Earth to Neptune is 1 AU × 10.03 days/365.26 days/(0.2754/360) = 35.9 AU.

But we are being premature. We can and should account for the fact that Neptune creeps slowly across the sky as it too orbits the Sun. This corresponds to a steady movement across the sky in exactly the opposite direction to that generated by the Earth's motion around the Sun. Neptune therefore moves through an angle of 360 / 164.8 / 365.26 = 0.00598 degrees per day. This must be added to the angle we computed above, and means that the correct parallax angle is (0.2754 + 0.0598) = 0.3352 degrees. With this correction, we can use the underlined formula to deduce that the distance to Neptune was 29.49 AU, which means that, since the planets were aligned, the distance from the Sun to Neptune was 30.49 AU.

It is possible to do a little bit better than this, and account for the fact that Neptune was not quite in opposition on 19 August 2014 because the Earth was not yet in line. Doing the trigonometry to account for this changes the 30.49 AU to 30.33 AU. We do not do it, but the optimal use of Kevin's photos would also make use of the other two measurements. If you want to do so then be aware that 21 September 2014 is over 3 weeks after opposition, so the trigonometry is not as straightforward.

Parallax measurements are so fundamental that astronomers have created a measure of distance directly related to them. If a star has a parallax of 1 arcsecond, it is defined as being 1 parsec away. These angular shifts are very small, which is why we do not perceive a shift in the shape of the constellations. Aristotle used the lack of a perceived parallax shift of the stars to argue for a stationary Earth; in large part his logic was sound, but he underestimated the distances involved and instead of saying that 'no parallax implies the Earth is stationary', he ought to have considered the alternative hypothesis, that 'no observed parallax means that the stars are a long way away'. To put the smallness of stellar parallax angles into context, the angular size of the full Moon is approximately 2000 arcseconds and the nearest star to Earth, Proxima Centauri, exhibits a parallax of just 0.762 arcseconds, which means it is $1/0.762 = 1.31$ parsecs away; 1 parsec is equal to 3.26 light years, which is the distance that light travels in 3.26 years.[3] Even the nearest stars are mind-bogglingly far away.

With modern technology, parallax measurements as small as 10 millionths of one arcsecond can be made. The most accurate measurements to date are being made by the European Space Agency's Gaia satellite. It was launched into orbit around the Sun in December 2013 from French Guiana, the location from which Jean Richer helped to make the first accurate parallax measurement of Mars almost 350 years earlier. Gaia's precision translates into parallax measurements out to distances of tens of thousands of light years,

---

[3] A star 1 parsec away is in fact $149.6 \times 10^6/(1/3600 \times \pi/180)$ km $= 3.1 \times 10^{13}$ km away. Light travels at 3×10^8 m/s, so this distance is also equal to 3.26 light years.

bringing a large fraction of the stars in the Milky Way within our reach.

When we consider distances to galaxies lying beyond our own Milky Way, we'll have to deal with distances measured in megaparsecs, or Mpc. A megaparsec is a million parsecs: 3.26 million light years. The minuscule parallax angles associated with such distant objects are impossible to measure, so if we want to map the Universe beyond the Milky Way, we need to develop other methods. Even so, it remains the case that all distance measurements in astronomy ultimately rely on parallax methods – if not directly, then to calibrate them.

On moonless nights away from lights, the sky is ablaze with stars, and every one visible to the naked eye is inside our galaxy. There are so many worlds in the rich starfields sweeping through the constellations of the galactic plane – Cassiopeia, Perseus, Sagittarius – that the light from a billion suns merges into a continuous glowing arch that, when viewed from the true darkness and still air of high mountain or desert, delivers the sensation of standing alone on the edge of a galaxy. The Milky Way is vast beyond imagination. 'Somewhere, something incredible is waiting to be known,' wrote Carl Sagan.

Close to the 'W' of Cassiopeia, in the constellation of Andromeda, the unaided eye can, just, catch a faint glimpse of a misty patch that lies beyond; it is the Andromeda galaxy, our closest large galactic neighbour. There are people alive today who were born before this was known.

Until the morning after the night of 5 October 1923, it was possible to take the view that our home galaxy constituted the entire Universe, and that those misty patches were structures inside the Milky Way. We can date with precision the moment when the last vestiges of our pre-Copernican centrality were kicked away, because on that evening, at Mount Wilson in California, the American astronomer Edwin Hubble took a 45-minute exposure of the Andromeda Nebula with the 100-inch Hooker telescope. In the now-famous photograph, shown in Figure 4.5, Hubble recognized three bright stars that had not been present in his previous photographs. He assumed that they were novae – bright flares from small, dense white dwarf stars caused by the accretion of matter from an orbiting stellar companion – and marked them on the photographic plate with the letter 'N'. The following day, checking through the observatory's archive photographs of Andromeda to confirm his result, he noticed that one of the novae was sometimes present and sometimes not; it appeared and disappeared with a period of approximately thirty-one days. He immediately realized that this wasn't a nova, but a type of star known as a Cepheid variable, which changes its brightness like clockwork. He excitedly crossed out the 'N' on his photograph and wrote 'VAR!' next to the star.

To understand why Hubble was excited enough to use an exclamation mark, we must step back a further fifteen years. In 1908, an astronomer at the Harvard College Observatory called Henrietta Leavitt published a paper with her fellow astronomer Edward Charles Pickering, in which she reported a relationship between the brightness of Cepheid variable stars sitting in the Small Magellanic Cloud[4] and their period. The period is the length of time over which the star's bright-

---

[4] It is now known to be a satellite galaxy of the Milky Way, some 199,000 light years from Earth.

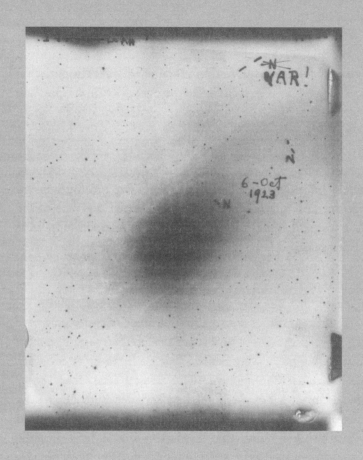

Figure 4.5 The photographic plate taken by Edwin Hubble
at Mount Wilson Observatory on 5 October 1923.

ness varies, i.e. the time from bright to dull back to bright again. Leavitt expressed the relationship in simple, precise prose: 'It is worthy of note that in Table VI the brighter variables have longer periods. It is also noticeable that those having the longest periods appear to be as regular in their variations as those which pass through their changes in a day or two.' Leavitt's observation became the basis of what is now known as the period–luminosity relation. The key observation is that the period of a variable star is indicative of its inherent brightness.

To understand why this was one of the most important breakthroughs in the history of astronomy, recall our problem with the measurement of the distances to very remote objects: their parallax shifts are too small to be detectable. This looks like a very serious problem at first sight. Stars are points of light with no discernible structure, and the distant ones do not appear to move. All we have is their light. The most obvious difference between stars is their brightness. If one star is brighter than another, two obvious reasons spring to mind: the star may actually be brighter, or it may be inherently less bright but closer to us. To simplify the logic, imagine if all stars were of the same intrinsic brightness: we could then use the observed brightness of a star to determine how far away it is relative to the other stars. For example, a star that was twice as far away as another would be ¼ as bright.[5] Now, if we used parallax to determine the

[5] This is because the brightness is a measure of the light power, and this falls away with the square of the distance from the source. For example, the power incident on a 1 cm^2 light detector placed 1 metre from a 10-watt light source will equal 10 watts × 1 cm^2/(4π × 1 m^2). This is just a statement of the fact that (to a good approximation) the light bulb emits light equally in every direction and the detector receives its share of the total.

distance to just one nearby star then we could determine the distance to any other star by comparing its brightness to the nearby star.

The trouble is that the inherent brightness of a star typically depends on its mass and age. Sirius is the brightest star in the night sky. It is only 8.6 light years away, but it is twice as massive and 25 times more luminous than our Sun. Rigel, in the constellation of Orion, is the seventh brightest star in the night sky, but it is around 860 light years away. Rigel is a supergiant star, with a diameter around a hundred times that of the Sun, and it is 200,000 times more luminous. The most luminous known star is in the Large Magellanic Cloud. It is a Wolf-Rayet star known as RMC 136a1. It shines with the brightness of 8.7 million suns, and is 315 times more massive. It can be seen with a small telescope, even though it is 163,000 light years away. So you see the problem with using observed brightness as a proxy for distance.

The importance of Leavitt's period–luminosity relation is that it sidesteps this problem by providing a sample of stars with known relative brightness. This is so important that it is worth spelling it out in some detail. Imagine that there are two Cepheids with precisely the same period. According to Leavitt, they must also have the same intrinsic brightness. Now suppose that one of them is four times as bright as the other, as seen from Earth. Immediately, we can conclude that the brighter star must be half as far away as the dimmer star. Using Leavitt's period–luminosity relation, we can therefore determine the distance to any Cepheid variable star in the sky, including the one Hubble identified in Andromeda, *provided* that we know the distance to at least one of them.

But we already know how this can be done: we should look

for nearby Cepheids whose distance can be determined by parallax. In 1913 the Danish astronomer Ejnar Hertzsprung made the first measurement of the parallax of a Cepheid variable. The star delta Cephei, from which Cepheid variables take their name, has a parallax of 3.77 milli-arcseconds, which puts it at a distance of approximately 890 light years (this is the modern measurement). It has a period of 5.366341 days. This is the 'standard candle', the star that Hertzsprung used to calibrate Leavitt's ruler.

(A brief historical aside: scientists are only human. In his original paper, Hertzsprung got the parallax to delta Cephei correct, but made a simple error in his estimate of the distance to the Small Magellanic Cloud using Leavitt's relation, which he quoted as only 3000 light years. This was a trivial mistake, which was quickly noticed, but for some reason Hertzsprung, one of the greatest modern astronomers, never bothered to correct it in the literature.)

Edwin Hubble knew all about these measurements on the morning of 6 October 1923 – hence his excitedly scribbled 'VAR!' His discovery of a Cepheid in the Andromeda nebula allowed him to determine the distance, which he calculated to be a shockingly large 900,000 light years. Modern measurements of the distance to Andromeda, based in part on a better understanding of Cepheid variable stars,[6] put the giant spiral at an even greater distance of 2.54 million light years. Historically, the factor of nearly three makes little difference because, whichever way you look at it, Hubble had determined that Andromeda sits well outside of the

---

[6] It was realized in the 1940s that there are two main classes of Cepheid, with different period–luminosity relationships. Not knowing this led to some confusion in the early days.

Milky Way (which is only 100,000 light years in diameter). He had shown conclusively that Andromeda could not be a nebula – a cloud of stars, gas and dust inside our galaxy. It is a separate island of stars, so distant that the light reaching us now, which you may be able to glimpse tonight with a decent pair of binoculars from your back garden, began its journey long before there was such a thing as a human being on planet Earth.

If you do venture out tonight with your binoculars, you may be able to see other nearby galaxies; M81 and M82 in Ursa Major are a good start, at a distance of around 12 million light years, which we know thanks to Cepheids.[7]

More powerful telescopes reveal greater numbers of galaxies. Plate 5 shows the Hubble eXtreme Deep Field (XDF) image, a very long-exposure photograph taken by the Hubble Space Telescope during 2003 and 2004. It corresponds to a tiny piece of sky (approximately 1/30th of one millionth of the entire sky), and still it contains over 5500 visible galaxies. The XDF photograph contains a wealth of information: data about a deep slice of space and time. It is a record of the light that fell on Hubble's mirror, virtually all of which originated from galaxies beyond our own.

Now we have the distance to nearby galaxies, using Leavitt's distance–luminosity relation for Cepheid variable stars, let's do a quick back-of-the-envelope calculation based on the number of galaxies visible in the Hubble XDF image. If the Hubble XDF is typical of the entire sky – and we have no reason to think otherwise – then there are approximately

---

[7] A Type 1A supernova was observed in M82 in January 2014, which makes for an accurate determination of how far it is away (see below).

30,000,000 × 5500 = 165 billion galaxies in the observable Universe. If we further assume that the average distance between galaxies is the same as the distance between the Milky Way and Andromeda, then we can estimate the size of the observable Universe to be $(165 \times 10^9)^{1/3} \times 2.5$ million light years = 14 billion light years. This is obviously a very rough estimate,[8] but it is one that gives us a sense of the scale of the Universe.

Since light takes 1 billion years to travel 1 billion light years, our estimate implies that the Universe is probably at least 14 billion years old, because the light from the most distant galaxies must have had sufficient time to make it across the Universe and onto Hubble's mirror. Simply by counting stars and by measuring the distance to Andromeda, we are led to contemplate a Universe that is old enough to contain the Sun and Earth, which we have already dated to over 4 billion years in age. Obviously the Universe must be older than the Sun and Earth, but it is nice to see how very simple reasoning leads to the correct ordering of their ages.

There are only a handful of measurable properties of the most distant galaxies, because astronomers have only the light these galaxies emit to work with. For a relatively nearby galaxy, we can determine its size if we know its angular size on the sky and how far away it is (which we can do, if we've been able to find a Cepheid variable star or some other object of known brightness). For the more distant galaxies, however, resolving individual stars is not possible. This means that the Cepheid method will not be any use for galaxies that

---

[8] It does ignore the expansion associated with the Big Bang, of which there will be more in the next two chapters.

are too far away. For these we need another new trick.

There are certain types of astronomical events, known as Type 1A supernova, that are very bright and can be seen at very large distances. To say they are very bright is something of an understatement, because a single Type 1A supernova shines brighter than entire galaxies of stars, albeit for just a few fleeting days. This type of supernova occurs when a star gains mass by consuming matter from a nearby star. Once the star becomes 40% more massive than the Sun it starts to collapse in on itself under the weight of its own gravity.[9] The fact that all Type 1A supernovae happen in essentially the same way means that they are all of the same intrinsic brightness. Just as with Cepheid variables, these can therefore be used as 'standard candles' to determine how far they (and their host galaxies) are away. The trouble is that they are rare events: the supernova rate in an average galaxy is estimated to be around one per century, and the last Type 1A supernova to be clearly observed in the Milky Way was way back in 1604. Many more have been seen beyond the Milky Way, which is why they are a very valuable tool for measuring cosmic distances.

There are other ways to measure the distances to galaxies that do not rely on the rare good fortune of observing a Type 1A supernova. We will meet two of these in Chapter 6, and use them to help us do some cosmology. Both methods exploit the fact that there is extra information encoded in the light from a galaxy, beyond the size and shape of its image on a photograph. There is also colour: not just the reds, blues and greens visible in photographs, but the precise, fine details

[9] We computed this 40% figure in our earlier book, *The Quantum Universe*.

of the distribution of the colours in the spectrum. This might not seem like much, but it is.

In Box 7 (pp. 107–9), we give a brief explanation of the way light is emitted and absorbed by atoms, which should help in understanding the last part of this chapter. For those who don't fancy reading the Box, the executive summary is that the light from any star or galaxy contains a character-istic barcode which tells us precisely what that star or galaxy is made of. This barcode is universal: the chemical elements that make up the stars and dust in the Andromeda galaxy are the same as the chemical elements that make up your body. The barcode of the Sun's atmosphere, taken during a solar eclipse, is shown in Plate 7. This spectrum was made by passing sunlight through a prism, or to be more precise a high-precision kind of prism known as a diffraction grating. The peaks in the spectrum signal the presence of different chemical elements in the Sun's atmos-phere: hydrogen, helium, calcium, magnesium, iron and traces of other elements can be seen. This is obviously a very neat way to ascertain what distant stars and galaxies are made of.

Now let's look at a galaxy. We've chosen a beautiful one called NGC4535, a barred spiral galaxy in the constellation of Virgo. The photograph in Plate 8 was taken using the 200-inch Hale telescope at the Palomar Observatory in California. The telescope was completed in 1948, and remained the world's largest until 1993. If you know a little about photography or engineering, you'll appreciate what a superb instrument this is when we tell you that at the prime focus of the 20 m^2 mirror the focal length is 16.76 metres with an aperture of f3.3. The Hubble Space Telescope has also observed several Cepheid variable stars in NGC4535, and from these we know the distance to be 52 million light years.

Just as for the Sun, we can examine the spectrum of light from NGC4535. A portion of that spectrum is shown in Figure 4.6, where we can see five strong emission lines. The strongest is at a wavelength of 6606 ångstroms. Because galaxies are expected to be predominantly hydrogen gas, we might suppose that this line is associated with hydrogen atoms. Indeed, in Earth-bound laboratory measurements there is an emission line close to this wavelength, called the H-alpha line, but on Earth the wavelength is 6563 ångstroms. This important line can also be seen in the spectrum in Plate 9 of Box 7. The Earth-bound and galactic wavelengths do not quite agree. Should we doubt our interpretation? Is the bright emission line from NGC4535 coming from something else?

To help understand what is going on, let's look at the two lines that lie on either side of the proposed galactic H-alpha line. They have wavelengths of 6591 and 6627 ångstroms. They are also close to two similar lines that are familiar from Earth-based experiments: a pair of emission lines from nitrogen known as the NII lines, with lab wavelengths of 6548 and 6584 ångstroms. Now we have a clue: all of these three spectral lines have wavelengths that are around 0.65% bigger than their earthly counterparts. This systematic shift of wavelengths is confirmed when we study the two smaller lines at 6760 and 6775 ångstroms. After accounting for the 0.65% shift, these correspond to two Earth-bound lines known as the SII lines, which are emitted by sulphur atoms, at 6716 and 6731 ångstroms. So, the spectrum from

NGC4535 looks exactly like a terrestrial spectrum – except that all of the wavelengths have been increased by the same factor of 0.65%.

We could conclude that there is something strange about galaxy NGC4535. Could the structure of atoms be different there? It's hard to see why or how this could be the case. The mystery deepens when we look at the spectra from other galaxies. The barcode patterns from *all* galactic spectra correspond precisely with the barcode patterns of atoms detected on Earth, but they are always shifted. Furthermore, they are all shifted by different amounts – but the over-whelming majority are shifted to higher wavelengths, just like NGC4535. This shift to higher wavelengths is known as a redshift.

Astronomers quantify the amount of redshift by the ratio of the shift in wavelength divided by the wavelength of the spectral line as it would be measured on Earth. This means that NGC4535 has a redshift of 0.0065. Redshifts can be much larger than this, though. In Figure 4.7 we show the spectrum of light from 3C273, a type of active galaxy known as a quasar. The redshift in this case is far greater, which you can immediately see by just looking at the plot. In this case, the bright H-alpha line is somewhere closer to 7600 ångstroms and the redshift is 0.1583. This is confirmed by the other lines (we can see several other hydrogen emission lines). As in the case of NGC4535, we see a barcode pattern just like on Earth – but this time with a 15.8% shift to higher wavelengths.

There are two possible ways to explain redshift. One is that atoms are different everywhere in the Universe, and vir-tually all of them emit light with longer wavelengths than

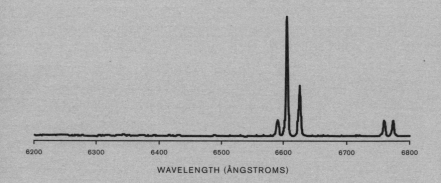

WAVELENGTH (ÅNGSTROMS)

Fig 4.6 The emission spectrum of NGC4535. The biggest bump is due to the presence of hydrogen and the two smaller bumps either side are due to nitrogen. The two bumps on the far right identify the presence of sulphur.

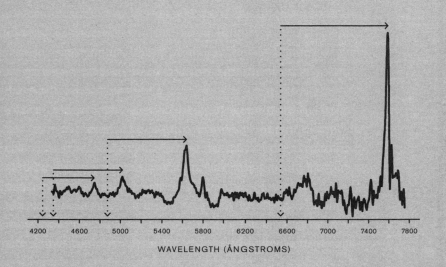

Fig 4.7 The emission spectrum of 3C273, which is a quasar at
redshift of 0.1583. This spectrum was obtained using the Hubble
Space Telescope. The arrows indicate the extent of the redshift.

those in our galaxy. The other possibility is that something happened to the light on its journey from the galaxies to our telescopes that caused the wavelengths to increase.

It is one of the most remarkable facts in the history of science that physicists were in possession of a theory that could explain the redshift of the galaxies years before Hubble had even confirmed that galaxies existed, and years before any redshifts were observed. That theory is Einstein's theory of General Relativity, which many physicists regard as the most beautiful physical theory of them all.

BOX 7. WHAT IS LIGHT?   P. 107

Light can be thought of as a wave. Like a water wave it has peaks and troughs. But unlike water waves, where we can see the peaks and troughs very clearly, the peaks and troughs in a light wave correspond to variations in the size of electric and magnetic fields, which is rather more abstract. We do sometimes feel the effect of these variations because, if the wavelength is right, the electric fields push the electrons around in our eyes to generate the signals that our brains turn into images. The distance between two peaks in a light wave is called the wavelength, and we interpret different wavelengths of visible light as different colours. We are able to see light with wavelengths between around 400 nm and 700 nm. Beyond our visual range, past the short-wavelength violet light, there are X-rays and gamma rays. Beyond the longer-wavelength reds lie the radio waves. The electromagnetic spectrum is summarized in Plate 9.

Light is emitted by atoms when they are heated up, and absorbed by atoms when it shines on them. From studies here on Earth, we know that each kind of atom – each chemical element – emits or absorbs only very particular wavelengths of light. Each element has a distinct signature, which is ultimately down to its individual atomic structure. Plate 10 shows the spectrum of light emitted and absorbed by hydrogen atoms. The absorption spectrum is like a rainbow with pieces missing, and it is made in much the same way, by shining white light through hydrogen gas and then through something like a prism. The prism acts like raindrops, splitting light into its component colours by spreading them out. The dark vertical lines across the rainbow in the spectrum shown in Plate 10 are known as absorption lines, and they are produced when light of a particular colour is absorbed by the hydrogen atoms. The emission spectrum is produced when we look at the light emitted by a hot gas of hydrogen.

The emission and absorption lines are at exactly the same wavelengths because emission and absorption are a result of electrons jumping around inside hydrogen atoms. Quantum theory explains why the electrons are only allowed to have specific energies when they are confined inside

BOX 7. WHAT IS LIGHT?   P. 108

atoms, and therefore why the emission and absorption lines are discrete. We don't need to get into the details of quantum theory here, but it is helpful to know that light can also be thought of as a stream of particles called photons, and that every time an electron in an atom loses energy, a photon is emitted from nowhere with energy exactly equal to that lost by the electron. These are the photons we see in the emission spectrum. Likewise, an electron can raise its energy by absorbing a photon of just the right energy (the photon then disappears). This leads to the absence of photons we observe in the absorption spectrum. The energy of the photons is inversely proportional to the wavelength of the light, which means that higher energy photons correspond to shorter wavelengths.

The pattern of emission and absorption lines is always very distinctive: it resembles a DNA fingerprint or, as we've seen, a barcode. This allows us to analyse the light captured by a telescope from a star or galaxy and identify which chemical elements are present. Plate 11 shows the light from the Sun, split up into its component colours. The hundreds of black absorption lines crisscrossing the rainbow are the atomic fingerprints, and this is how we know precisely what the Sun is made of, even though nobody has ever been there.

Good science is often about paying attention to the smallest, seemingly insignificant details of Nature, and there is one detail we can't resist mentioning. Take a look again at Plate 7, which shows the emission spectrum of the outermost parts of the Sun (called the chromosphere and corona) observed during a solar eclipse. The eclipse helps us to observe an emission spectrum, because the Moon blocks the background light from the rest of the Sun. This is the light emitted by atoms in the solar atmosphere. The emission lines in Plate 7 can be paired with lines of exactly the same wavelengths in the solar absorption spectrum, which we showed in Plate 11. This is to be expected. The absorption spectrum, which is the thing we usually see, is created when the white light from the surface of the Sun shines through the solar atmosphere. The chemical elements then absorb light according to their characteristic colours, creating the

BOX 7. WHAT IS LIGHT? P. 109

barcode. As we have already noted, if the atoms are present to absorb light, then they are also present to emit it, but we don't usually see the emission spectrum because the bright glow of the hot surface overwhelms it.

However, if you look closely, you can see something odd. The two lines marked He 5876 ångstroms and He 4472 ångstroms are not visible in the solar absorption spectrum. The former is the second most intense line in the emission spectrum, which means it must be abundant in the outer portions of the Sun. These lines are due to the presence of the element helium. Helium, the second most abundant element in the Universe, was not discovered on Earth but by studying the emission spectrum from the Sun. During the solar eclipse of 1868, the English astronomer Joseph Norman Lockyer saw the two bright emission lines, which at that time corresponded to no known terrestrial element. Appropriately enough, the new element was christened after Helios, the Greek sun god: helium.

If, as we now know, the Sun is 27% helium outside of the core, why do we not see it in the much more visible solar absorption spectrum? The reason is this: the Sun's surface mainly glows at a relatively cool 6000 degrees celsius, at which temperature helium is transparent to visible light, so, even though there are helium atoms present, no absorption line is seen. It is one of the most intriguing facets of our star the Sun that its outermost atmosphere is far hotter than its surface, which is a result of the complex behaviour of its magnetic field. The corona has a temperature of around 3 million degrees celsius, which is definitely hot enough to excite the electrons inside helium atoms and make them emit light. Unless the much more intense light coming from the bulk of the Sun is blocked out (as it is in an eclipse), these emitted photons are too few to be easily noticed. This is why helium is present in the emission spectrum, but not in the absorption spectrum.

5.
EINSTEIN'S THEORY
OF GRAVITY

All things fall with the same acceleration under the influence of gravity. That statement is pretty well known and it doesn't sound very profound, but it is. In terms of an equation, you may recall that the acceleration of an object falling to the ground is given by $a = GM/r^2$. Here, G is the strength of the gravitational force so skillfully pinned down by Henry Cavendish in the late eighteenth century, r is the distance from the centre of the Earth, and M is the mass of the Earth. Nowhere in this equation is the mass of the falling object present, which is the reason for the opening sentence to this chapter. Big, heavy things, like the Moon, accelerate to the Earth at the same rate as light things, like a mote of dust. This is the famous result obtained by Galileo, which he is said – probably apocryphally – to have demonstrated by dropping two balls with different masses from the Leaning Tower of Pisa. Newton used this result to argue that the force of gravity must be proportional to the mass that appears in his Second Law of Motion, $F = ma$, which states that if you apply a force to something, it accelerates by an amount inversely proportional to its mass. The m in $F = ma$ is known as the inertial mass, because it describes how hard it is to push something. It is because of the precise equivalence between *gravitational* mass, used in Newton's Law of Gravitation, and *inertial* mass, used in Newton's Second Law of Motion, that the acceleration due to gravity is independent of the mass of the falling object. As we shall now see, the equivalence of gravitational and inertial mass has profound consequences.

Newton had an answer to why these two conceptually different masses happen to be equal: they are equal because that is the way things are. Nobody did any better than that for over two centuries, until a young Albert Einstein realized that, as he put it, 'for an observer freely falling from the roof of a house, at least in his immediate surroundings, there exists *no* gravitational field'. Einstein appreciated that freely falling objects feel no force, which means they do not accelerate: this is why the paths they follow are independent of their mass. Looking back in 1920, Einstein described this as 'the happiest thought of my life'. To understand precisely why he said this – Einstein was not prone to exaggeration – you need to understand what follows in this chapter. An unexpected spin-off of this happy thought, which undoubtedly surprised Einstein as much as everyone else, was the prediction that there may well have been, to use George Lemaître's poetic phrase, a day without a yesterday. In other words, Einstein's theory of gravity predicts the Big Bang.

Let's think carefully about free-fall. Imagine jumping off a roof. As you fall, according to Einstein, it is as if somebody switches gravity off. That sounds odd. Hurtling towards the ground, it would be a brave soul who dares to claim there is no gravity.

In the twenty-first century, we don't have to imagine this situation, because we are all familiar with images of astronauts aboard the International Space Station (ISS). From the perspective of an observer on the ground, the astronauts are falling, together with the space station and everything aboard, towards the ground due to the gravitational pull of the Earth.

The astronauts are, however, floating. It is a common misconception to imagine that the astronauts are floating in 'zero-G' because they are a long way from Earth; they are not. The altitude of the ISS is only 400 km, and the radius of the Earth is 6370 km. That corresponds to a 10% reduction in the gravitational pull of the Earth relative to that felt by a person falling from a roof at the Earth's surface. The space station doesn't hit the ground when it falls because it's also travelling tangentially at a velocity of 7.66 km/s, which means that it is continually missing the ground. It's precisely the same for the Moon. It too is forever falling to the Earth and forever missing, as Newton well knew. This is what being in orbit means. Watching the footage of the astronauts, it's obvious that gravity at least *appears* to have gone away. If an astronaut lets go of a cup, or a hammer, or even a globule of water, it stays put. A water globule doesn't move relative to the astronaut, or the space station, or fall towards the Earth faster or slower, despite the fact that its mass is relatively tiny. It just floats. This is what we call 'zero-G' – but there is, according to Newton, plenty of 'G' present.

Einstein and Newton describe this behaviour in completely different ways. Newton puts the equivalence between inertial and gravitational mass centre stage, and this is his explanation for the fact that nothing moves relative to anything else on the ISS. Einstein simply asserts that nothing moves relative to anything else because there is no force acting on anything.

Hopefully you can now see what Einstein is driving at. In a sense, the astronauts aboard the ISS can't tell whether they are falling towards the Earth or simply floating, far away from any gravitational influence in interstellar space. If we close off the option of looking outside, there is no experiment they can perform or observation they can make that will inform them either way. In which case, asserts Einstein, there really is no difference between the two situations. People in free-fall do not experience gravity because there isn't any: 'for an observer freely falling ... at least in his immediate surroundings, there exists *no* gravitational field.' If we describe the world from the point of view of people in free-fall, the force of gravity will never enter into things.

But surely something more needs saying: after all, gravity really does exist. If you jump off a roof, you do hit the ground. Einstein has an answer to this. Yes, the distance between you and the ground decreases to zero when you jump off a roof, but that is because the ground is accelerating up to meet you. Again, you may feel uncomfortable with this statement, but you have to admit that this is what being on the ground actually feels like. As you sit reading this book, you are being not-so-gently pressed into your chair. It feels the same as when you are pressed back into the seat of an aircraft when you accelerate down the runway. This is because it *is* the same, according to Einstein. Your weight isn't anything to do with the force of gravity as Newton understood it. Rather, it's the feeling you get because you are accelerating upwards as a result of the force exerted on you by the ground, to meet the poor unsuspecting sod who's minding his own business floating around after stepping off a roof.

Surely not, you may say. Imagine two people falling freely at two different places on the Earth; perhaps one is above England and the other is above Australia. Following Einstein, they could each legitimately claim to be floating freely in space whilst the ground is rushing up towards them. The ground cannot be accelerating towards each of them at the same time, you might say, because this would imply the Earth is getting bigger. But that logic is wrong.

Imagine two small balls, separated by some distance, falling freely towards the Earth. From Newton's perspective, each ball accelerates towards the centre of the Earth because the force of gravity acts along a straight line drawn between the centre of the ball and the centre of the Earth. This means that the two balls must get closer together as they fall, because they each fall along straight lines that meet at the centre of the Earth. If the balls start only a metre apart, and fall over a distance of only a few metres, this is a tiny effect. But imagine separating the balls by a few thousand kilometres, and dropping them from a similar altitude. They will close in on each other by an appreciable amount as they fall, as illustrated in Figure 5.1. The most extreme example would be to separate the balls so widely that one falls over England and the other over Australia, in which case they would hurtle towards each other. Einstein's theory must be capable of explaining this, and of course it can. So how can two balls get closer together when they are experiencing no force? The explanation is the key to understanding General Relativity.

We've met Newton's Second Law, $F = ma$ (the force acting on an object equals that object's mass times its acceleration), but not his first. Newton's First Law states that 'every body remains in a state of rest or uniform motion in a straight line unless acted upon by a force'. We have said that our falling balls experience no force, because we've dispensed with gravity, and yet they get closer to each other as they fall in the vicinity of the Earth. They are moving together without a force. This seems like a contradiction, but we've chosen our words carefully. Here is an example of 'moving together without a force': imagine two explorers standing a small distance apart on the equator, and imagine they agree to journey due north, walking in perfect straight lines at constant speed. There are no net forces acting, so they will continue to move in perfect straight lines in accord with Newton's First Law. But they will get closer together and bump into each other at the North Pole. They move closer together because straight lines on the surface of the Earth are lines of longitude, which cross the equator at right angles, but converge on each other towards the pole. There is no force pulling the explorers together; they get closer to each other as they move north because they are moving in straight lines across the curved surface of the Earth. The geometry of the space over which they are moving is not flat, and because of this they are drawn closer together as they move.

This is why the two balls dropped above England and Australia move towards each other as determined by someone standing on the surface of the Earth. They move towards each other just as the two explorers move towards each other on their way to the North Pole. Each freely falling ball (or explorer) can 'claim' quite legitimately that the other is accelerating towards them, while they feel no acceleration. Now we can understand why it is possible for freely falling observers in both England and Australia to each claim that the ground is accelerating up to meet them, and yet the Earth is not getting any bigger. It is because acceleration is not universal. To be explicit, we can go back to the example of our two explorers. Imagine they walk along paths that pass either side of Iceland. According to each explorer, they personally are not accelerating, because they are moving along straight lines, but Iceland is getting closer to each of them as they journey past it on their way to the pole. Iceland, they each claim, must be accelerating towards them, in the sense that it is moving with a changing speed and direction relative to them. Iceland is accelerating from the perspective of both explorers, but it obviously doesn't change shape in reality.

According to Einstein, Newton's force of gravity is as fictitious as the 'force' pulling those two explorers together. We are misled into suspecting the presence of an attractive force that causes things to fall when, in reality, they are moving in straight lines over a curved surface. But what is the surface in the case of objects falling in space? This is where it becomes harder to visualize what is going on in our mind's eye because the surface is not a surface of two dimensions, like the surface of the Earth. In Einstein's theory, the 'surface' is both space and time. The mathematically simple but intuitively tricky idea of a surface in more than two dimensions is explored in Box 8 (pp. 119–20).

Let's summarize what we have so far, because this is counter-intuitive. Just as people walking due north walk 'over the Earth' and get drawn together 'because it is curved', so two balls in free-fall move 'over space and time' and get drawn together 'because it is curved'. The 'it' we refer to is a 'surface' made from space and time, a.k.a. spacetime. The force of gravity is a fictitious force; it is a manifestation of the fact that spacetime is curved.

The word 'genius' is certainly overused and, in science, it gives the impression that progress is the result of a series of eureka moments experienced by a handful of intellectual

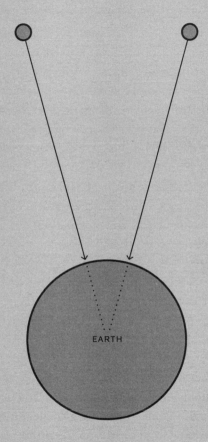

Figure 5.1 Two balls falling towards the Earth.

freaks. This is not usually the case. Science is a collective endeavour. Having said that, if it ever was appropriate to use the word, then Einstein's development of General Relativity must qualify. It took Einstein a great deal of time and effort to arrive at his explanation for the force of gravity. He had the idea that there are no forces acting on an object in free-fall sometime in late 1907; it was late 1915 when, having worked out the consequences for a new theory of gravity, he published them as his General Theory of Relativity.

The beauty of Einstein's General Theory of Relativity is in part due to the simplicity of its core idea: that objects move over a curved spacetime in straight lines. The curvature of spacetime does all the work of what used to be called 'force'. In this way, the Earth and the Moon are bound together in orbit, and yet both travel in perfect straight lines over spacetime. It is so brilliant and enchanting an idea that many modern theoretical physicists hope that all of the forces of Nature can be dispensed with and replaced by geometry, using General Relativity as the template.

The part of General Relativity that tells us how things move over curved spacetime is only one piece of the theory. The other, very necessary, part tells us how spacetime becomes curved in the first place. Obviously, in the case of an object falling near the Earth, it must be that the Earth is responsible for warping the spacetime around it. More precisely, and more generally, the presence of matter and energy curves spacetime. It is pretty obvious that this has to be the case, because gravity is something we naturally associate with planets and stars, and bigger things exert more gravitational pull than little ones (think of astronauts leaping on the Moon). General Relativity contains a set of equations known

BOX 8. HIGHER DIMENSIONAL SPACE   P. 119

When we speak of spacetime as a 'surface', we are using the word in a mathematical sense, in order to make a parallel with the two-dimensional example of the Earth's surface apparent. But spacetime is four-dimensional. It has three dimensions of space plus the one dimension of time. Do not get alarmed by the fact that the surface we are speaking of is four-dimensional and that one of those dimensions is time. You will certainly not be able to picture this; nobody can. Fortunately, Nature is not restricted to things that human beings can picture: Nature is richer than that. And, also fortunately, human beings have discovered mathematics, which allows them to deduce things that they cannot picture.

In mathematics, if we're allowed a 2-dimensional surface, then we're also allowed a 3, 4, 5 or n-dimensional surface. All we need to do is add more numbers in order to specify the co-ordinates of points that lie on the surface; we also need to specify some way of calculating distances on the surface. You can picture the surface of a ball without any problem, but in an imaginary flat-world, where there is no 'up' and 'down', the inhabitants would really struggle. They could still do mathematics concerning the surfaces of balls, though. They might speak of a ball as the mysterious two-dimensional surface that generalizes the one-dimensional notion of a circle (a circle is something they would be able to picture). Figure 5.2 shows a picture of a four-dimensional cube – well not quite. It is the shadow cast in three dimensions by a four-dimensional cube, drawn in projection in two dimensions (i.e. on a sheet of paper). As you can see, we're able to get a glimpse of what higher dimensional objects look like by looking at the shadow they cast in the lower dimensions, just as your shadow would be something that a flat-world inhabitant could visualize.

BOX 8. HIGHER DIMENSIONAL SPACE   P. 120

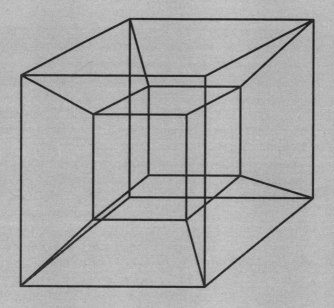

Figure 5.2 The shadow of a tesseract.

as the Einstein Field Equations, which provide the mathematical means to determine the shape of spacetime if we know the way that matter and energy are spread about. Here they are:[1]

$$R_{\mu\nu} - \tfrac{1}{2}Rg_{\mu\nu} + \Lambda g_{\mu\nu} = \frac{8\pi G}{c^4}T_{\mu\nu}$$

Einstein's equations may look foreboding – but, like many of the equations that describe how Nature behaves, they tell us stories. The term $T_{\mu\nu}$ on the right contains the details of the matter and energy distribution in space and time. If we wanted to use Einstein's equations to describe the solar system, we'd put a spherical blob in here with the mass of the Sun. The terms on the left describe the resulting geometry of spacetime, which tells planets how to move in the vicinity of the Sun. As an aside, the symbol Λ is known as the cosmological constant, and we'll meet it later on. You will recognize at least one symbol: G, Newton's gravitational constant, which describes the strength of gravity. In Einstein's equations, it tells us how much spacetime curves in response to a given distribution of matter and energy.

We can use Einstein's equations to compute the warping of spacetime close to the Earth's surface. This gives results that are almost identical to those predicted by Newton. We say 'almost', because there are some differences. For example, Einstein's equations predict that time passes more quickly at the top of a mountain than at the bottom; remember, both space and time are warped by the presence of the

---

[1] Although this looks like only one equation, the Greek subscripts can vary (they label projections in space and time), giving rise to several distinct equations.

Earth, and not just space. This effect is so tiny that it would probably have remained undetected until the twenty-first century had people not bothered to look for it, but it is a large enough effect that, today, it must be accounted for in technology that requires high-precision timing, such as the GPS system.

Things get more dramatic when we look beyond the Earth. Soon after Einstein published his theory in 1915, two of his predictions that differed from Newton's were tested. One concerned the orbit of Mercury, which is not well described by Newton's laws and had been a known problem for almost two centuries. Einstein's theory, it turned out, correctly predicted the observed motion of Mercury. The other was Einstein's prediction for the amount by which light is deflected in a gravitational field, which differed from naïve Newtonian-inspired predictions by perhaps the most famous factor of 2 in all of physics. Einstein's result was confirmed during the solar eclipse of 1919, when starlight could be observed passing close to the Sun, and this was the moment he became world-famous. Today, Einstein's theory has been tested to remarkable precision in a variety of astrophysical systems, systems often bizarre and violent beyond imagination.

The most spectacular recent triumph was the discovery of gravitational waves: ripples in the fabric of spacetime. At 5.51 a.m. EST, on 14 September 2015, the twin LIGO detectors, located in the United States in Washington State and Louisiana, detected gravitational waves as they passed through the Earth (see Plate 12). These waves were caused by the collision of two black holes, themselves also a prediction of Einstein's theory. The black holes, 29 and 36 times the mass of the Sun, spiralled into each other in less than two tenths of a second, during which time their closing speed changed from one third to almost two thirds of the speed of light. The collision resulted in a peak power output fifty times that of the entire observable Universe, which distorted space and time enough for the effects to be measured 1.3 billion light years away on Earth. Einstein's theory predicted with precision the observed signal. What more could you ask for?

So there we have it: Einstein's theory of gravity, General Relativity. It supersedes Newton's Law of Gravitation because it delivers a more accurate description of Nature. That is not to deny that Newton's law remains excellent in many instances, and has the advantage of being far easier to deal with mathematically. But the hard evidence indicates, without a shadow of doubt, that Einstein's theory is more correct than Newton's.

We can now be a little more ambitious. We've seen that Einstein's theory is extremely good at describing the curvature of spacetime around spherical blobs of matter, like planets, suns and even black holes. We might therefore be tempted to ask whether it could also be applied to a larger distribution of matter: the entire Universe, for instance. This is audacious to say the least – and, naturally, the thought had already occurred to Einstein. In 1917 he published a paper entitled 'Cosmological Considerations of the General Theory of Relativity', the audacity of which did not escape him. 'I have ... again perpetrated something about gravitation theory which somewhat exposes me to the danger of being confined in a madhouse,' he wrote in a letter to his friend Paul Ehrenfest. Let us, with due humility, follow in Einstein's footsteps. This will be well worth it, because what follows is a triumph. Einstein's theory *predicts* the existence of the redshifts of the galaxies that we encountered in the last chapter. And that is because it also predicts the existence of the Big Bang.

If we knew all about the distribution of matter and energy in the Universe, we could put that information into Einstein's equations and obtain the geometry of spacetime. We could then ask if there are any observable consequences of this geometry, and look to test the theory. This is what we will now do.

There is, of course, no way we can know where all the material in the Universe is located. It is hard enough to keep track of a pair of socks. We are not stymied, though. What follows is an example of good science. We are going to make a simplifying assumption. As ever, if our assumption is wrong, we expect to find out when we compare our predictions with experiments or observations of Nature.

Plate 13 shows a map of the galaxies visible from Earth, compiled from several databases by Thomas Jarrett, an astronomer at Caltech. It's worth looking for a moment at this picture, because it is remarkable and humbling. There are so many galaxies that for the most part only clusters and superclusters of galaxies are labelled. The Virgo Cluster alone contains over 1300 galaxies that lie around 53 million light years from Earth. A deep photograph of the cluster from the European Southern Observatory in Chile is shown in Plate 15, revealing a snowstorm of galaxies: delicate clouds of light from billions of worlds.

The sky is filled with galaxies, scattered in what looks like a random pattern. They appear uniformly distributed on the

BOX 9. GRAVITATIONAL WAVES   P. 125

Until the LIGO detection of gravitational waves, astronomers were limited to using electromagnetic waves to observe the cosmos – now they have a whole new way of seeing. The LIGO detectors are staggering in their sensitivity and they are the modern day equivalent of Henry Cavendish's experiment (see pp. 68–72). The two LIGO detectors are shown in Plate 14 and they each consist of two 4 km long high-vacuum 'arms' with heavy mirrors hanging at either end. The mirrors are suspended like pendula on fused silica fibres. You can see the long arms on the photos. As a gravitational wave passes, it pushes and pulls the mirrors so as to change the distance between them, and the scientists are able to measure the change in length of one arm with respect to the other (by arranging the arms at right-angles one arm tends to increase in length when the other decreases). They do this using a technique called laser interferometry, which involves sending laser light into each arm, where it bounces back and forth many times off the hanging mirrors before re-emerging. The strength of the combined output light is sensitive to variations in the roundtrip time for the light to travel along each arm – so if the lengths change then the output laser light will reflect that change.

Figure 5.3 shows the amount by which the difference in length between the two arms of the detector changed as the wave of September 2015 passed by. This is labelled 'strain', and it is the fractional amount by which the arms changed in length. Notice how minuscule the strain is. It means that the scientists were able to measure the change in length to better than 1 part in 10^{21}. That relative precision is like measuring the distance to our nearest star, Proxima Centauri, to the thickness of a human hair: it is mind-blowingly precise. In terms of absolute distance, it corresponds to measuring the change in distance between the mirrors to about one thousandth of the size of a single proton. You might well doubt that such a precision is possible, not least because the surface of the mirrors has variations that are much bigger than that. But the brilliance of the measurement sidesteps the problem because it measures an average distance. Each photon that enters

BOX 9. GRAVITATIONAL WAVES   P. 126

the apparatus is, in a sense, making its own measurement, and one of LIGO's key features is its ability to harness the power of vast numbers of photons. To understand how powerful this averaging is, imagine a totally fictitious wave that has the effect of making everyone on Earth increase in height by one hundredth of a millimetre. If you knew when the wave was due to pass, could you detect it? Well, you certainly couldn't tell with any confidence from measuring the height of any one person before and after the supposed passing of the wave, because you simply wouldn't be able to measure their height accurately enough. But you could do it if you measured the difference in height (before and after the wave passes) of everyone and then took the mean. That is because the uncertainty on the mean decreases as the number of people increases[1] and so, with enough people, you could spot whether or not their average height has increased slightly after the wave was due. This is similar to what is done to measure that minuscule change in the arm lengths in LIGO. The main limitations on the precision of the experiment are due to the limited number of photons they have at their disposal and to the small vibrations induced in the hanging mirrors due to seismic effects and the fact that a laser is bouncing off them. Passing trains, ocean waves and the weather all induce relatively large variations but they can all be eliminated to a sufficient degree with diligence and attention to detail. This is brilliant science straight out of the Henry Cavendish school.

Figure 5.3 not only shows how the strain changed with time over the fraction of a second when the wave was passing, it also shows the theoretical prediction of what the signal ought to look like if it came from a pair of colliding black holes. It is wonderful to see the extent of the agreement. The speeding up of the wiggles reflects the speeding up of the black holes as they spiralled towards each other, and the final part of the curve, when the wiggles

[1] The percentage error on the mean falls as $1/\sqrt{N}$ where N is the number of measurements.

BOX 9. GRAVITATIONAL WAVES   P. 127

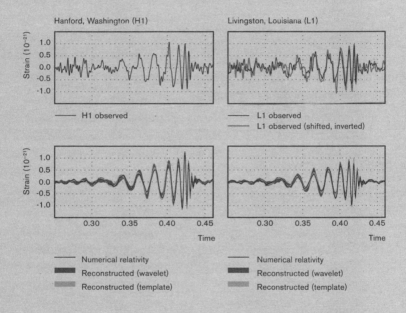

Figure 5.3 The all-important graphs demonstrating the discovery of gravitational waves. The two left-hand graphs show the measurement from Hanford and the corresponding theoretical prediction. The right-hand graphs are the same but for the detector at Livingston.

BOX 9. GRAVITATIONAL WAVES   P. 128

eventually die away, corresponds to the final phase of the merger, when the two black holes have become one.

A very exciting postscript to the story is that on 15 June 2016 the LIGO team announced their second detection of a pair of colliding black holes. The event occurred on Boxing Day 2015 and involved smaller black holes (8 and 14 times the mass of the Sun). Again, the observed signal agrees perfectly with the calculations based on Einstein's theory. A second observation so soon after the first is very encouraging because it means that it is likely there will be much more to see in the years to come, and that we are seeing these events for the first time now because we have finally managed to develop the technology to do so. This surely heralds a new dawn in astronomy.

largest distance scales, and this is the clue we need to make our assumption. We will suppose that the matter in the Universe is scattered evenly. This is of course not true over small distances: the solar system is very lumpy, with all the mass being concentrated in the Sun and a few planets. But we are going to assume that over 'sufficiently large' distances, the Universe is smooth. In other words, if we imagine counting the number of galaxies inside an imaginary box, sufficiently large so as to contain many galaxies, then that number will not vary much if we choose another equally sized box somewhere else in the Universe.

In the jargon, this is called a homogenous and isotropic matter distribution: the same everywhere and the same in every direction. It's the ultimate statement of the Copernican principle. We do not occupy a special place in the Universe, and neither does anyone else.

If we want to calculate the shape of all of spacetime, we know what to do. We plug our homogenous and isotropic matter distribution into the right side of Einstein's field equations, and see what comes out on the left side. Fortunately, we can immediately anticipate some of the results without actually doing any mathematics.

A perfectly smooth matter distribution throughout the Universe leads to a perfectly smooth spacetime. Galaxies, stars and planets will cause the spacetime in their vicinity to be distorted compared to the perfectly smooth average. Think of it like a golf ball, which we can picture as a perfectly smooth sphere with small dimples that constitute local, small distortions on the surface. The local dimpling of the golf ball is analogous to the way spacetime is 'dimpled' by local clustering of mass. The Sun will make a tiny dimple. A galaxy will make a bigger one. When we want to consider the large-scale structure of the Universe, we are not interested in the dimples. We are interested in the smooth spacetime that represents the Universe at large.[2]

We can now think about what a 'perfectly smooth spacetime' might look like. It turns out that there are only three possibilities, which we shall refer to as 'flat', 'spherical' and 'hyperbolic' (you need to know some mathematics to be able to prove that there are no more possibilities). Remember that spacetime is four-dimensional, with three spatial dimensions, so we can't picture it very easily. Fortunately, the same three possibilities also arise in two spatial dimensions, and they are sketched in Figure 5.4. We made the initial assumption that there are no special places and no special directions in the Universe. This must imply that the shapes of spacetime we get out of Einstein's equations have no special places and no special directions. It is easy to see that there are no special points or directions on a flat sheet. Likewise, there are no special points or directions on the surface of a sphere, the second of the three possible geometries. The third and final possibility is a saddle, which is rather like an inside-out sphere. The curvature of the surface is the same everywhere, but instead of being convex like a sphere, it is concave. The three possibilities in the four dimensions of spacetime are analogous to these, and the analogy is good enough for us to dare to draw some galaxies on the surfaces in the figure. These are the three possible universes under the assumptions of homogeneity and isotropy, and it is correct to think of them as showing how space is curved at some instant of time.

---

[2] We will get seriously interested in the dimples in Chapter 8.

It is impressive that we can make statements about the possible geometries of spacetime using such general reasoning, but we can go further. Look at Figure 5.5, which shows a hypothetical flat universe at three different times (we could equally well have chosen one of the other two geometries). The figure indicates that our model universe is being evenly stretched as time advances – just as one might stretch a rubber sheet by pulling its edges equally in all directions. Crucially, if the space is to remain flat, spherical or hyperbolic, then the only conceivable thing it can do as time advances is to shrink or expand. At this point in our narrative, we have no idea whether the space we live in expands, contracts or stays the same. Remember, Einstein has encouraged us to think of space as something malleable: the presence of matter and energy can and does distort it. This is the origin of planetary orbits and the reason things hit the ground when they are dropped. The idea that it can stretch as well as warp, in response to the matter and energy within it, should not be so bizarre.

Now is the time to ask what the full machinery of General Relativity has to say. We would like to know exactly how the Universe stretches or shrinks with time. Once we know that, we know everything about how space evolves in a perfectly uniform universe. Other questions we might have will be concerned with deviations from the 'same everywhere' assumption; we'll come to that in Chapter 8.

Imagine two points in the Universe that are a certain distance apart at some instant in time. Einstein's equations tell us how that distance changes as time advances. In other words, they tell us how the Universe is stretching or shrinking. Not surprisingly, the detailed answer depends

Figure 5.4 The three possible geometries in 2D,
with some galaxies drawn on.

TIME

Figure 5.5 The flat universe from the previous figure at three different degrees of stretching.

on which of the three geometries we live in and on what kind of stuff fills the Universe, but one conclusion is pretty hard to circumvent: in Einstein's theory of General Relativity, the distance does tend to change. This is extremely important, because it immediately implies that a Universe full of stars and galaxies like ours could well be expanding or contracting.

If you did not know any better, the idea of an expanding Universe is surprising, and one might ask whether the equations also allow for the possibility of a static universe. Einstein asked this question. In 1917 he showed that a universe without stretching or shrinking is (just about) conceivable, but to do it he had to introduce the cosmological constant, which we briefly mentioned when we met his field equations. Einstein included the cosmological constant simply because there is no obvious reason why it should be excluded, and even today there is no good understanding of what it might represent at a deeper level. One way to think about a cosmological constant is to regard it as a weird form of matter. Ordinary matter, of the type that makes up stars and planets and galaxies and people, is made up of particles. If we capture some of these particles in a box, they zip around and collide with the walls and exert a pressure that acts to try and expand the box. Einstein's weird matter can be thought of as doing the exact opposite: it would suck the walls of the box in rather than push against them. This might sound contrived, and in fact Einstein later dismissed the introduction of the cosmological constant as his 'greatest blunder'. Historians argue about what Einstein meant by this. One possibility is that he described it that way because, in spending so much time trying to force General Relativity to describe a static universe, he missed the fact that his equations were screaming out to him to entertain the idea of a Universe that is expanding. This uncharacteristic blindness meant he missed the chance to predict the Big Bang.

But two of Einstein's contemporaries, Alexander Friedmann in the Soviet Union and the Belgian Georges Lemaître, did not miss this message. Independently, they correctly deduced that Einstein's General Theory of Relativity, when applied to a Universe such as ours, predicts that it should be expanding or contracting.

Alexander Friedmann was the first to entertain the idea of an expanding[3] Universe. He used Einstein's equations to write down an equation for what is known as the scale factor of the Universe. The scale factor is a number that tells us how much space is stretched or contracted relative to its size today – in an expanding universe, the scale factor would be smaller than 1 in the past and bigger than 1 in the future. Used by cosmologists today, the equation is, unsurprisingly, called the Friedmann equation. We explore the equation in Box 10 (pp. 138–40). Friedmann made a calculation for the spherical geometry of spacetime in 1922 and for the hyperbolic geometry in 1924. He did not discuss the possibility of a flat geometry, nor did he comment on astronomical implications: rather, he was more interested in pointing out that Einstein's static solution (corresponding to a scale factor that does not change with time) was not necessarily correct. Lemaître re-discovered the Friedmann equation in 1927, although, in focusing on a spherical geometry he too missed the flat-space

[3] Although 'expanding Universe' is the common parlance, 'stretching Universe' is probably better.

solution. Lemaître's solution of 1927 described an infinitely old spherical universe with an ever-increasing scale factor.

Despite these decisive contributions, both Friedmann and Lemaître were not so well known in the scientific community, and their work was largely ignored when it was first published – partly because of the prevailing prejudice towards a static universe, supported by the venerable Einstein, but also for more mundane reasons. Lemaître's paper appeared in French in the relatively obscure *Annales de la Société Scientifique de Bruxelles*, and was not widely read; Friedmann, meanwhile, died of typhoid in 1925, shortly after publishing his work. In 1930, however, the well-known British physicist Arthur Eddington began to publicize Lemaître's work, when he came to realize that Einstein's static universe was implausible because even the tiniest deviation from the idealized conditions that Einstein postulated would cause the universe to expand or contract. The dust finally settled in 1933, when Princeton physicist Howard Percy Robertson wrote a beautiful paper that carefully catalogued all of the possible mathematical solutions to the Friedmann equation. He counted eighteen of them, spanning the various logical possibilities corresponding to flat, spherical or hyperbolic space and varying amounts of cosmological constant. Robertson's conclusion was that, using only Einstein's equations and pure logic, a universe containing matter can only be:

(i)   of a finite age and expanding for ever into the future;
(ii) of a finite age and expanding for some period of time before contracting back again;
(iii) of infinite age, for ever expanding (the expansion rate would need to tend towards zero in the distant past);

   (iv) of infinite age with no expansion or contraction;

   (v)   of infinite age, first contracting and then expanding.

In those solutions where the universe has a finite age, there must have been a time in the past when the universe was infinitely dense, meaning that the average distance between any two particles was zero. The equations start to fail when the distance between the particles becomes too small, but the notion that there was a time when the Universe was extremely densely packed with particles is a robust prediction. That special time in the Universe's history is what we now refer to as the Big Bang. Scenario (iii) is the solution found by Lemaître in 1927 and scenario (iv) is Einstein's static solution. Scenario (ii) is interesting because it has a 'Big Crunch' at some point in the future, and scenario (i) is interesting because, as we will see next, it is the Universe we actually live in.

BOX 10. THE FRIEDMANN EQUATION   P. 138

The Friedmann equation is without a doubt the most important equation in cosmology. Here it is:

$$H^2 = \frac{8\pi G}{3}\,\rho - K\,\frac{c^2}{a^2 R^2}$$

The equation tells us how fast space expands when the scale factor of the Universe is equal to a. We have written the equation so that the scale factor $a = 1$ at the present time. If the Universe is expanding, a was smaller than 1 in the past and it will be bigger than 1 in the future. The Friedmann equation tells us precisely how a changes as time changes. Let's go through each bit of the equation in turn. On the left of the equals sign is H, which is the Hubble rate: it tells us how fast space is expanding at some moment in time and it is equal to the fractional rate of change of the scale factor a. In Chapter 6, we will make our own measurement of the present day value of H, and we'll find it is equal to approximately 70 km/s/Mpc, which means that (today) space is expanding such that two objects that are 1 megaparsec apart are moving apart at a speed of 70 kilometres per second. Because space is assumed (on average) to be the same everywhere, this means that objects that are 2 megaparsecs apart are currently receding from each other at a rate of 140 kilometres per second, and so on. If we know how H has changed in the past, and if we know how it will change in the future, then we will also know how the scale factor a has changed in the past and how it will change in the future. In others words, we will know everything there is to know about the expansion history of the cosmos: one of the holy grails of cosmology.[1]

Now let's turn to the right of the equals sign. The first term depends on Newton's gravitational constant, G, and also on the average mass density of the Universe, denoted by ρ. This is the amount of mass in every cubic metre of space, averaged across the Universe. The second term

---

[1] If you know some elementary calculus then it's easier just to say that $H = \dot{a}/a$ where $\dot{a} = da/dt$ and then we obtain a as a function of the time t by square rooting each side of the equation and integrating.

BOX 10. THE FRIEDMANN EQUATION   P. 139

is where the geometry of space enters into things. The symbol K just keeps track of the sign: it is equal to +1 (corresponding to a spherical geometry) or −1 (hyperbolic geometry) or, if this term is absent, it is 0 (corresponding to a flat geometry). c is the speed of light and R is the curvature of the Universe, which needs specifying if the Universe is either spherical or hyperbolic, because we need to know how big the sphere or saddle is (a two-dimensional hyperbolic space looks like a saddle).

The equation therefore tells us that space is expanding at a rate that is fixed by how much stuff it contains, and its shape. This is the essence of Einstein's equation.

If you are keen to play about with the Friedmann equation then you will want to know how the average mass density depends on the scale factor, because clearly it will depend on it. If you are not keen then you can safely skip the rest of this paragraph. In the case of matter particles, like protons or dark matter, the number of particles per unit volume will decrease as space stretches. This means that the density is equal to the present-day density divided by a^3 (think of measuring the density by measuring how much matter is in a cube – and that the cube shrinks or grows by a factor a; its volume will correspondingly shrink or grow by a factor a^3). The case of light[2] is a little more subtle; it contributes to ρ by virtue of having energy, and the appropriate mass density is obtained by dividing the energy per unit volume by c^2. This is how Einstein's theory of gravity differs from Newton's: Einstein understood that both mass and energy contribute to gravity. You might think that the present-day contribution to ρ from light would also need to be divided by a^3 in order to determine the contribution at some other time. But in fact it needs to be divided by a^4. This is because light becomes redshifted as space expands (or blueshifted as it contracts), which means its energy density rises or falls in inverse proportion to the scale factor. The last remaining contribution to ρ comes from the energy that may

---

[2] Or any matter moving at speeds close to the speed of light, in fact.

BOX 10. THE FRIEDMANN EQUATION   P. 140

be stored in empty space, such as might be associated
with a cosmological constant. This source of mass density
does not change as the Universe expands or contracts,
so it contributes a constant value to ρ. Now you can go
ahead and solve the Friedmann equation using a computer
(the integral you need to do is too hard to do by hand): if
you know the present-day mass densities associated with
light, matter and the cosmological constant then the above
information is enough for you to compute the value of the
scale factor a for all time. We will determine the present-day
mass densities in Chapter 7.

6.
THE BIG
BANG

Part of what makes Einstein's General Relativity so astonishing is that it was, more or less, an exercise in pure thought, triggered by his musings on the universality of free-fall. It's wonderful that we can trace the development of Einstein's happiest thought all the way through to Robertson's classification of the possible histories of the Universe; to follow this simplest, most elegant of ideas into uncharted and unexpectedly rich terrain.

This, of course, is a book about the Universe we live in and not an imaginary one, however beautiful. Today, General Relativity has been tested to exquisite precision – the LIGO detection of gravitational waves that we discussed in the last chapter being the most recent and most striking example. Now, our goal is to establish the extent to which the Universe as a whole fits into Einstein's framework. We want to gather evidence in support of the idea that we live in an expanding Universe that started with a Big Bang.

Our first task is to establish whether the Universe is expanding now and to measure the rate of that expansion. We have already seen one piece of evidence that indicates our Universe is expanding: the way that light from distant galaxies is redshifted. According to General Relativity, this occurs because the expansion of the Universe causes successive crests and troughs of light waves to move apart. This means that the wavelength of light arriving at the Milky Way from a distant galaxy should be increased (i.e. redshifted) by the same fraction that the distance between the two galaxies has increased during the time the light took to make the journey. Space stretches, and the light stretches with it. If the redshift is due to the expansion of the Universe, then we can predict that light from the most distant galaxies should have the biggest redshifts, because the light has been travelling for longer, meaning that space will have stretched more. Let's see if these features are borne out, by taking a look at some real data.

We'll use the NASA Extragalactic Database, which is freely available on the web, and explore how the redshifts of a sample of galaxies vary with their distances from Earth. So far in the book, we have tried to make our own measurements, so looking in a database might sound like cheating, but it is only a little cheat. It is possible for an amateur astronomer with a reasonably sized telescope costing a few thousand pounds, a spectrograph, a digital imaging system and a laptop to make galaxy redshift measurements for galaxies out as far as 100 Mpc. The absurdly suspicious or commendably enthusiastic reader who doesn't trust the databases is encouraged to take this route. Indeed, if for some reason you don't believe the redshift data, then you *should* take this route, because it will inform you that your opinion is wrong and the databases are right. Taking delight in being shown to be wrong is one of the most important skills any human being, let alone a scientist, should develop.

In order to test how redshift depends on how far away a galaxy is, we need precise distance measurements to far-away galaxies. These are more difficult for an amateur to perform. As we have seen, the Cepheid method can be used for galaxies close enough for individual stars to be resolved and, for more distant galaxies, supernovae can be used, but only for those galaxies that happen to contain a supernova that we are lucky enough to have seen. To make a comprehensive map of the distances to the galaxies, we really could do with another way to measure how far away distant galaxies are.

There are several other widely used methods of measuring the distances to galaxies. One of the most accurate was developed in 1977 by North American astronomers R. Brent Tully and Rick Fisher. The Tully-Fisher method can be used to measure distances to spiral galaxies, which are so named because of their beautiful spiral shapes. The Milky Way and Andromeda are both spiral galaxies – in Plate 16 we show a photo of Andromeda; it is a disc-like assembly of stars that is rotating about an axis through its centre. Tully and Fisher's method uses the measured brightness of a spiral galaxy and the 'width' of its spectral lines[1] to determine how far away it is, and it is explained in Box 11 (pp. 148–149). In the spirit of this book, however, we would also like a simple 'Ogmore-by-Sea' method that will allow us to check that the more sophisticated distance measurements make sense.

Here is our simple method. Since all spiral galaxies look quite similar to each other, we will be so bold as to assert that

[1] Recall that the spectral lines emitted by a galaxy correspond to the barcode pattern of light they emit. Each line corresponds to a particular atomic transition that leads to the emission or absorption of light of one particular wavelength. Each spectral line is strongly peaked at that particular wavelength, but it does still have a 'width', which means there is a small spread of wavelengths about the peak value. You can see this small spread in the spectra we showed in the plots at the end of Chapter 4 (the spikes are not super-sharp). This width contains important information, as Tully and Fisher realized.

they are all roughly the same size. Under this assumption, we can estimate the distances to all of them if we know the distance to just one. For example, if a galaxy is half the size of the Andromeda galaxy, as viewed from Earth, we'll assume it is twice as far away, and so on. It may or may not be a good approximation to say that all spiral galaxies are exactly the same size, but we might reasonably expect the variation in sizes to be not so great as to spoil things. In any case, since we have the accurate distance measurements to hand in the NASA database, we can check whether our method stands up under the illuminating spotlight of the precision data. The angular size of the Andromeda galaxy on the sky is 6200 arcseconds,[2] and the distance to Andromeda from Earth, obtained from Cepheid variable measurements, is 780,000 parsecs, or 2.5 million light years. Our old friend NGC4535 has an angular size of 410 arcseconds and, if we assume it's the same size as Andromeda, this means it must be 15 times further away, which puts it at a distance of 12 Mpc. In total, we've selected sixteen spiral galaxies at random from the NASA database; the distances and redshifts to these galaxies are listed in Table 6.1. Photographs of the galaxies are shown on Plate 17. There is nothing special about these sixteen galaxies except that they are all spirals at redshifts between 0.001 and 0.02.

We chose the upper limit of 0.02 because we wanted to be able to identify the galaxies as spirals by eye, which

---

[2] It is not entirely obvious how to measure the size of a galaxy, because you have to decide precisely how the edge of the galaxy is defined, and which wavelengths of light are used to make the measurement. We use the so-called '2MASS' measurements, which are made in the near-infrared part of the spectrum. For our purposes, it doesn't matter which method we choose, as long as we are consistent and use the same method for every galaxy. For more details, see http://iopscience.iop.org/article/10.1086/345794/pdf.

allows us to use the 'Ogmore-by-Sea' method to estimate their distances from the Milky Way. As we've seen, a larger redshift means a greater distance, and therefore a smaller galaxy in the sky. If you look at Plate 17, you'll see that our highest redshift galaxy, UGC6533, is just about identifiable as a spiral, but we might struggle to identify spirals at

Galaxy	Redshift	Angular size (arcseconds)	Estimated distance (Mpc)	Measured distance (Mpc)
NGC5055	0.00164	694	7.1	8.3 ± 1.8
NGC7171	0.009069	164	30	41 ± 4
UGC00465	0.0159	88	56	66 ± 3
NGC1073	0.00403	230	22	14 ± 2
NGC1313	0.00157	440	11	4
NGC3887	0.00403	217	23	19 ± 2
NGC3684	0.00388	126	39	24 ± 2
NGC0011	0.0146	127	39	55 ± 4
NGC7280	0.00615	138	36	23 ± 2
NGC0150	0.00528	113	43	20 ± 3
UGC6533	0.0179	84	59	84 ± 4
NGC0251	0.0152	108	45	64 ± 3
NGC0337	0.00549	197	25	22 ± 2
NGC4535	0.00655	410	12	17 ± 3
NGC4712	0.0146	146	34	74 ± 9
NGC3668	0.0117	107	46	62 ± 5

Table 6.1 A list of sixteen spiral galaxies along with their redshifts, angular sizes in the sky and the distances away from Earth. The column marked 'Measured distance' is the best estimate of professional astronomers (mainly using the Tully-Fisher relation). The column marked 'Estimated distance' is our simple estimate that assumes all spiral galaxies are the same size. The measured distance is quoted with an uncertainty, which tells us about the accuracy of the measurement.

greater distances and, as a result, we might bias ourselves by selecting bigger than average spirals that are easier to see.

There are also spiral galaxies with redshifts smaller than 0.001. Indeed, our closest neighbour Andromeda has a negative redshift of −0.001. This means that the light is shifted towards the blue part of the spectrum, signifying that Andromeda is hurtling towards us and will hit us in around 4 billion years. This blueshift of Andromeda is caused by the Doppler effect, which we explained in Box 11 (pp. 148–149). We can understand the blueshift of Andromeda once we appreciate that galaxies are not only carried along with the expansion of the Universe. They also move around because of local gravitational interactions, just as the Earth orbits the Sun and the Sun orbits the centre of the Milky Way. So, we have to factor in this 'proper motion' of a galaxy, as well as its motion due to the expansion of space: accounting for it would lead to an additional red or blue Doppler shift superimposed on the redshift caused by the expansion of the Universe. As far as the Andromeda galaxy is concerned, the blueshift completely dominates, because it is so close that the effect of the expanding Universe on its light is very small. For sufficiently distant galaxies, however, the cosmological redshifts are large enough to overwhelm any Doppler shifts caused by their proper motions – which is why we only selected galaxies at distances above 4 Mpc, corresponding to redshifts above 0.001.

The approximate distances to the other galaxies in Table 6.1 are worked out in the same way that we worked out the distance to NGC4535 and appear in the column labelled 'Estimated distance'. These are to be compared with the numbers in the 'Measured distance' column, which were obtained by the professionals.

The closest galaxy in the list is NGC1313, at a distance of around 4 Mpc, and a redshift of 0.00157. The most distant is UGC6533, whose redshift is a factor of 10 larger at 0.0179. Looking at the table, the most distant galaxy exhibits the largest redshift, and the other galaxies appear to be such that larger redshifts are associated with more distant galaxies. This is in line with what we expect for an expanding Universe. To make our observation more visible, we should plot the data on a graph, and this is what we've done in Figure 6.1.

Immediately, the pattern in the data springs into view: all the points are scattered around a straight line. In Figure 6.2 we show Edwin Hubble's variant on this type of graph, which he produced in 1929. Hubble's vertical axis is slightly different from ours, because he chose to present his results in terms of the 'recessional velocity' of the galaxies rather than the redshift. As we will see in a moment, the recessional velocity is equal to the redshift multiplied by the speed of light, so Hubble's data extend to redshifts as large as 0.003 (1000 km/s divided by 3×10^5 km/s).

Our graphs and Hubble's have straight lines drawn through the data points. These correspond to the 'best-fit' lines, which means they are the optimal lines that can be drawn through the data. The data from the sixteen galaxies are scattered in the vicinity of the line but, because the measured distances to the galaxies are uncertain, the data are consistent with the line being the true description of what is happening. By this we mean that, if we are confident in the redshift value of a galaxy, then the line will give us a better measurement of the distance than the Tully-Fisher measure-

BOX 11. THE TULLY-FISHER RELATION   P. 148

The Tully-Fisher relation can be used to determine how far away a distant spiral galaxy is. It exploits a correlation between how fast a spiral galaxy is spinning on its axis and how bright it is. According to Newton's Law of Gravitation, the speed at which a distant star will orbit around the centre of a galaxy is determined by the mass of the galaxy, see Box 5 (p. 65). This makes sense, because more massive galaxies exert a bigger gravitational pull, which means stars can circle around the galaxy at higher speeds without flying off into space. But the wavelength of the light emitted from a star and observed on Earth depends partly on how fast the star is moving relative to the Earth. Specifically, if the star is moving towards the Earth the light will be slightly blueshifted and if it is moving away from the Earth the light will be slightly redshifted. This change in colour (wavelength) is not due to the expansion of space. Rather, it arises for much the same reason that a high-pitched siren on a fast-moving ambulance drops to a lower pitch as the ambulance passes by. This is called the Doppler effect, and it comes about because the sound waves emitted by the siren are compressed (shorter wavelength) when the ambulance approaches and stretched out (longer wavelength) when the ambulance recedes.

BOX 11. THE TULLY-FISHER RELATION   P. 149

Because light is also a wave motion, there is a Doppler effect for light. In the case of a spiral galaxy, this small shift in wavelength means that the spectral lines we observe on Earth are not precisely fixed at one particular wavelength: they are spread out a little bit, with some light being redshifted and some blueshifted relative to the average. The size of the spread is determined by how fast the stars move and, as we just said, this is correlated to the mass of the galaxy. In addition to this, the amount of light a galaxy emits depends on its mass, which is pretty obvious, because more massive galaxies contain more stars. Since the Doppler broadening of the spectral lines and the amount light a galaxy emits both depend on the mass of the galaxy, it follows that there must be a correlation between the amount of Doppler broadening and the amount of light that a galaxy emits, i.e. faster-spinning galaxies will have broader spectral lines and they will also be more massive. This correlation can be used to determine the distance to a galaxy by measuring the galaxy's brightness on Earth (more distant galaxies will be dimmer), provided that we first calibrate the method by measuring the distances to a few spiral galaxies using an independent measurement, such as the Cepheid variable method.

ment does. For example, the table shows that NGC0011 has a redshift of 0.0146 and its measured distance is 55 Mpc. But from the left-hand graph in Figure 6.1, you can use the line to conclude that this distance is probably a bit too low, and that the actual value might be more like 65 Mpc. To do that you need to notice that the data point corresponding to NGC0011 is the sixth point from the right.[3]

The line in the left-hand graph corresponds to the equation $z = 2.23 \times 10^{-4} \, \mathrm{Mpc}^{-1} \times d$, where d is the distance and z is the redshift. Notice that this allows us to determine the distance to any astronomical object simply by observing its redshift – that's to say, take the redshift and divide it by 2.23×10^{-4} to determine the distance in megaparsecs (this gives 65.5 for NGC0011). This technique provides astronomers with yet another way to determine distances to far-away objects. Strictly speaking, we should only use the equation for objects whose redshifts lie between 0.001 and 0.02, because we haven't plotted any data outside this range.[4] The number $2.23 \times 10^{-4} \, \mathrm{Mpc}^{-1}$ is called the gradient, or slope, of the graph. This is a very important number because, apart from calibrating the redshift–distance relation, it tells us how fast space is expand-

---

[3] Don't be concerned that the quoted uncertainty for the distance to NGC0011 is only 4 Mpc; a 10 Mpc error is only 2.5 times the uncertainty, which is not so unlikely that it should worry us (especially since there is also some uncertainty on the best-fit line). There is a well-established procedure for understanding how to interpret the uncertainties on measurements, but we do not propose to go into it here.

[4] In fact, if we went out to higher redshifts (around 1.0 and bigger), the equation eventually fails because the points no longer fall in the vicinity of a straight line. Even then, astronomers can still use redshift to ascertain distance. It would take us too far off track to discuss here exactly how they do that – suffice to say that if we know the average material composition of the Universe, which we will by the end of the next chapter, the Friedmann equation allows us to tell how the scale factor of the Universe has varied with time. This information is sufficient to determine the distance to any object whose redshift is known.

ing. Let's now understand why it should do so.

For a galaxy that is a distance d away, the light takes a time d/c to reach Earth, where c is the speed of light. During this interval of time, the Universe has stretched by the same factor as the light, which is the redshift z. This means that the galaxy has receded from Earth by a distance zd in a time d/c, i.e. the galaxy is receding from the Earth at a speed $v = zd/(d/c) = cz$. This is the recessional velocity that Hubble plotted on the vertical axis instead of the redshift. Hubble's graph demonstrates what is now known as Hubble's Law, which states that the recessional velocity of a galaxy is $v = Hd$, where H is equal to the slope of Hubble's plot. Naturally enough, H is known as the Hubble constant, and it is one of the most important numbers in physics because, as we will see in a moment, it is the number that tells us how fast space is stretching. According to Hubble's original data, H had a value of approximately 500 km/s/Mpc (something you can read directly off his graph). According to our graph, H is equal to the speed of light multiplied by the slope of our line, i.e. 3×10^5 km/s $\times 2.23 \times 10^{-4}$ Mpc^{-1} = 67 km/s/Mpc. Clearly our result is not compatible with Hubble's. The reason for the discrepancy is that Hubble underestimated the distances to his galaxies, just as he did when he correctly identified that Andromeda is far outside the Milky Way. Notwithstanding this, his plot was the first experimental evidence to indicate that the Universe is expanding.

A Hubble constant of 67 km/s/Mpc means that a galaxy that is 1 Mpc away from Earth is moving away from the Earth at a speed of 67 km/s due to the expansion of space (and a galaxy at 10 Mpc is moving away at 670 km/s). This linear relationship between redshift (or recessional velocity)

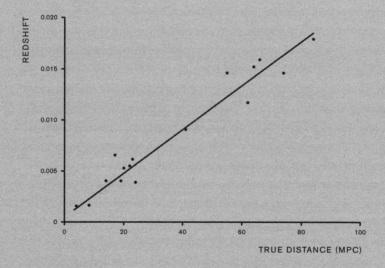

Figure 6.1 The redshift of the spiral galaxies in our sample, plotted against their distances from Earth. The graph on the left is the Hubble plot made using distances as determined by professional astronomers, and the graph on the right is the Hubble plot made using estimated distances to the galaxies under the assumption that they are the same size.

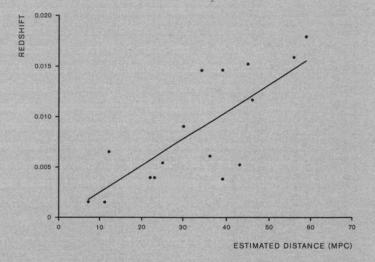

and distance is precisely what we would expect if space has been expanding at a constant rate. This means our spiral galaxy data confirm that space has been expanding at a uniform rate, at least for the past 300 million years (because we only looked at galaxies out to 100 Mpc, and the light travel time from these galaxies is just over 300 million years).

A nice way to visualize how Hubble's law comes about is to draw lots of spots on the surface of a balloon and then inflate the balloon. The dots all rush away from each other as the fabric of the balloon stretches; the dots that are further apart rush away from each other faster. If the expansion rate of the balloon stays constant, the balloon Hubble constant will be measured to be the same for each dot, as seen from any other dot, and we'll get a 'straight line balloon Hubble plot'. Substitute 'galaxy' for 'dot' and this is what we see in our spiral galaxy data. The surface of a balloon is a two-dimensional example; a three-dimensional example is the case of a cake containing raisins being baked inside an oven. As the cake expands, the raisins move away from each other. The Big Bang would then be like the moment when we start to cook a huge (possibly infinitely big) blob of dense dough. So, our Universe is more like baking a cake than inflating a balloon. This illuminates one of the more common misunderstandings in modern cosmology. The observational fact that all galaxies beyond our nearest neighbours appear to be rushing away from us does not mean we are at the centre of the Universe, in the same way that no dot on the surface of a balloon can be said to be at the balloon's centre. Rather, the data lend support to the picture, which stems from General Relativity, in which the galaxies are riding along in a Universe whose space is expanding.

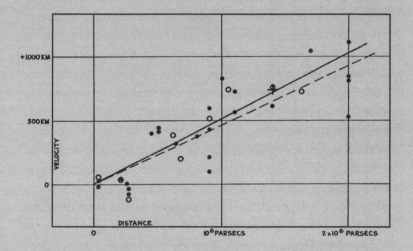

Figure 6.2 Edwin Hubble's original graph. The
velocities with which nearby galaxies recede from
Earth constitute the y-axis and their distances
from Earth are shown on the x-axis. (The woefully
labelled y-axis should read km/s, not km.)

To finish off this analysis, let's take another look at the right-hand graph in Figure 6.1, which is made using our simple estimate of distances to galaxies. Using it, we obtain a Hubble constant $H = 79$ km/s/Mpc. This is satisfyingly close to the more accurate result, and while the data are far more scattered around the line than for the left-hand graph, the general trend towards higher redshifts as the distances increase is still apparent. The broad agreement between the two approaches indicates that our assumption that all spiral galaxies are roughly the same size was not too bad after all. It is satisfying when a simple approach is able to produce an approximate result that corroborates a more sophisticated analysis – it is a kind of sanity check. It is even more satisfying when we learn that the most precise determination of the Hubble constant, made using data collected by the European Space Agency's Planck satellite, is 67.8 ± 0.9 km/s/Mpc. This is terrific physics: anybody, standing in their back garden with a reasonably sized amateur telescope and a few thousand pounds worth of kit, can prove that we live in an expanding Universe and measure the rate of the Universe's expansion.[5]

Let's recap for a moment. We have used observational data on sixteen spiral galaxies at distances between 4 and 84 Mpc to determine that the Universe has been expanding at a constant rate for the past 300 million years or so. We know this because the data on our Hubble plot fall on a straight line. This fits with our expectations based on General Relativity. While we used a database rather than measurements we made ourselves, we hope that it doesn't bother you too much

---

[5] All of the sixteen spiral galaxies we selected can be observed and measured by a keen amateur.

by this stage. As our exploration of the Universe gets more ambitious, we must inevitably reach the point where the measurements we need to make require more than just a camera and a map of Ogmore-by-Sea. If this weren't the case, we wouldn't need to spend billions of pounds on sophisticated telescopes and Large Hadron Colliders. The age of the lone experimental scientist is almost over, certainly in particle physics and cosmology, because all the simple measurements have been made, and there is admittedly a certain loss of romance in this. At the end of the eighteenth century, Wordsworth wrote of the statue of Newton at the entrance to Trinity College, Cambridge:

> And from my pillow, looking forth by light
> Of moon or favouring stars. I could behold
> The Antechapel where the Statue stood
> Of Newton, with his prism and his silent face,
> The marble index of a Mind for ever
> Voyaging through strange seas of Thought, alone.

Nevertheless, the romantic deficit is not total. It is still open to us to understand, or even to participate in, the large experimental collaborations that have replaced the individual in gathering precise data about the Universe. If we spend some time understanding how the data were acquired, and satisfy ourselves that we understand the measurements, there is no reason why we shouldn't use the data to voyage alone through strange seas of thought and to convince ourselves of the veracity of the remarkable picture of reality that modern science delivers. So, from now on, we will rely heavily on the data collected by modern-day teams of astronomers as we head rapidly towards the frontiers of current understanding.

We are still quite some way from establishing the existence of the Big Bang. That's because, although we have managed to confirm that the Universe has been expanding at the same rate for at least the last 300 million years, such a confirmation only rules out scenario (iv) from Robertson's list (p. 137): the static, eternal Universe initially favoured by Einstein himself. It's time now to turn to the evidence that the Universe has been expanding for nearly 14 billion years.

We are going to do this by supposing there actually was a Big Bang and then exploring what the consequences of that supposition might be. Of course, it is pretty well known that there is a good deal of evidence in support of the Big Bang – but the process of making a guess and seeing where the guess takes us is something that underpins how cutting-edge science is often done. As the iconic theoretical physicist Richard Feynman put it, when describing the search for new laws of Nature:

In general we look for a new law by the following process. First we guess it. Then we compute the consequences of the guess to see what would be implied if this law that we guessed is right. Then we compare the result of the computation to nature, with experiment or experience, compare it directly with observation, to see if it works. If it disagrees with experiment it is wrong. In that simple statement is the key to science. It does not make any difference how beautiful your guess is. It does not make any difference how smart you are, who made the guess, or what his name is – if it disagrees with experiment it is wrong. That is all there is to it.

Here, our guess is that we live in an expanding Universe of finite age, which is to say that there was once a Big Bang. In other words, we are guessing that our Universe is either scenario (i) (expand for ever into the future) or scenario (ii) (expand and then contract to a Big Crunch) on Robertson's list. Our guess will force us to make a prediction for the relative amounts of hydrogen and helium in the Universe, and we can then compare the result of the computation to Nature, as Feynman put it, to see if it works.

If our Universe started with a Big Bang and has been expanding ever since, there would have been a time in the distant past when any two points were very close together. The particles that make up the most distant stars would have been within a centimetre of the particles that make up your body. Of course, they would once have been even closer than a centimetre – but that is already mind-boggling enough. The point is that, in a universe that has been expanding for its entire lifetime, there will have been a time when things were squashed together into a state of very high density. Imagine squashing all of the planets in the solar system into a region the size of a pea, never mind all the stars in the Milky Way galaxy and all of the galaxies in the sky. As far as particle physicists can ascertain, the particles that make up atoms are of no discernible size, so the notion that all the matter in the observable Universe was once contained within a tiny region is not as crazy as it sounds. In any case, we don't need to let our imagination run quite so wild for the time being. We just need to suppose that at some time, long ago, the Universe was of a much higher density than it is today. That idea alone is going to prove very fruitful.

Incidentally, saying that the entire observable Universe was once compressed into a tiny region of space is not the same as saying that, at the time of the Big Bang, the Universe burst forth from a tiny region of space. The word 'observable' in 'observable Universe' is crucial in making this distinction. The observable Universe is the collection of things that we know to exist because we can see them.[6] There may be more to the Universe than this, but we can't see the other stuff because light travels at 'only' 300,000 kilometres per second and hasn't had time to reach us from very distant objects. We therefore do not know how big the Universe is – and it may be infinitely big, in which case it would have been infinitely big at the time of the Big Bang. To visualize this, we can return to the cake-baking analogy. Just after the Big Bang, the particles were close together; as time passes, the 'cake' stretches and the particles (like the raisins) move away from each other. This analogy invites us to picture the Big Bang as something that happened everywhere in space, and perhaps invites us to rename it the Big Stretch. What's more, there is no reason why the 'cake' couldn't have started out infinitely big: it can still expand and the particles will still be observed to move away from each other as time passes.

Now let's focus our attention on the physical conditions when the Universe was very much denser than it is today, and all the particles were very close together. We can immediately say that it was very hot. The idea that a gas heats up when it is compressed into a smaller region of space is familiar to anyone who has used a bicycle pump to inflate a tyre. Let's imagine a journey back in time. The clumping of gas

[6] By this we mean we can know of their existence by using some clever measuring device and our intellects. Eyes are only one of a range of 'seeing' devices available to scientists. So the dark matter counts as being part of the observable Universe.

and rocks into stars and planets and the clustering of stars into galaxies will be undone. If we go far enough back, even atoms – which are clumps of protons, neutrons and electrons – will break apart as they melt in the searing temperatures. In fact, let's go back to a time when the Universe was a hot, dense, featureless gas of elementary particles; protons, neutrons, photons, electrons and neutrinos. The temperature is now around 1 billion degrees, and the Universe is only a matter of seconds old.[7] It is likely that there were some dark matter particles too, but we know that these interact very weakly with ordinary matter, meaning that they were not major players in the drama we are about to describe. If you don't buy that, which is fair enough because we haven't presented any evidence yet, then we can guess that the dark matter is passive and see what predictions emerge.

We have chosen to wind the clock back to when the Universe was a billion degrees because we are interested in the question of the origin and abundance of chemical elements like helium in the Universe, and this is the time when the first atomic elements were assembled from the primordial protons and neutrons. Although we are guessing that such conditions were once present in the Universe (a hypothesis suggested by our prior observation that the Universe is currently expanding), we most certainly do not have to guess about the physics that takes place at such temperatures.

---

[7] To figure this out we need to solve the Friedmann equation for the scale factor and use measurements like those we describe in the next chapter to quantify how many particles of each type there are per unit volume at the present time. Saying the Universe is 1 second old really means that 1 second earlier all of the particles would be directly on top of each other unless some new physics steps in to change things. As we will see in Chapter 8, perhaps something did step in to change things, but at times earlier than we are considering in this chapter.

Everything that we are about to describe is very well tried and tested nuclear physics.

Isolated neutrons do not live for very long. One of them has a typical lifetime of about 10 minutes, after which it decays into a proton, an electron and a neutrino. The word 'isolated' is important, because neutrons can live for ever if they bind together with protons to make atomic nuclei. In the dense, 1-billion-degree-hot soup, a neutron can collide with a proton before it has time to decay and the two will stick together through the action of the strong nuclear force to form a deuterium nucleus. If the soup was much hotter, the deuterium would quickly break up again, but at temperatures below 1 billion degrees it is stable. Deuterium nuclei can then fuse with more protons and other deuterium nuclei to form helium and, eventually, very small amounts of lithium.

Around 3 minutes after the Big Bang, as the Universe expanded and cooled, the temperatures fell below those at which such nuclear fusion reactions can occur, and the synthesis of the first elements stopped. This process is known as Big Bang Nucleosynthesis. Textbook nuclear physics calculations predict what it produced: approximately 25% by mass helium and 75% hydrogen with traces of deuterium and lithium, the amounts of which can be calculated provided we know the ratio of the number of photons to the total number of protons and neutrons in the Universe. Turning this round, the calculations allow us to determine the photon-to-proton ratio, if we can measure the amount of deuterium.

The fact that the Sun contains roughly 27% helium, excluding the helium that has been produced in its core, is an encouraging start. However, the task of making precise comparisons between the nuclear physics calculations and the astronomical observations is not entirely straightforward, mainly because the stars fuse hydrogen in their cores to make heavier elements – and, when stars run out of fuel and explode, these newly minted nuclei can get scattered around, polluting the conditions that pertained after Big Bang Nucleosynthesis. To circumvent the uncertainties that arise from not knowing precisely how much pollution there is from dead stars, astronomers identify regions in the Universe in which the action of stars has not had a large effect. One way to do this is to look at the abundance of elements such as oxygen, which could only have been made in stars. If very little oxygen is present in some region of space, we can infer that the products of the nuclear reactions inside stars have not yet polluted the local environment. There are places known as HII regions (pronounced 'H two'), particularly inside dwarf galaxies, which satisfy this criterion. Astronomers can also look at very distant, bright objects known as quasars, some of which are over 10 billion light years away. We are seeing these objects as they were less than 4 billion years after the Big Bang, and this is not sufficient time for much stellar nucleosynthesis, so the quasars exist in a more pristine environment. Just as predicted, these regions contain 25% helium and 75% hydrogen.

The most accurate measurements of the deuterium content of the early Universe come from these quasars, and they tell us that there are around 30 deuterium nuclei for every million hydrogen nuclei. Primordial elemental abundances can also be measured by observing the light emitted from the outer layers of very old stars, which would reflect the composition of the primordial gas clouds out of which they formed. There are stars in the Milky Way galaxy that are well over 10 billion years old. Some of these, known as 'lithium-plateau' stars, are used to measure the lithium content of the early Universe. There is around one lithium nucleus for every 10 billion hydrogen nuclei in these stars' outer layers.

The good agreement between the observed and predicted relative amounts of primordial helium and hydrogen provides our first piece of evidence that the Universe has been expanding, not only for the past 300 million years (which is what our Hubble plot demonstrated), but at least since it was a hot plasma at a temperature of a billion degrees (a plasma is a hot gas of electrically charged particles). The theoretical calculations are also able to explain the primordial abundances of deuterium and lithium.[8] We said that these abundances can be predicted only if we know the photon-to-proton ratio: a ratio of 1.7 billion photons for every proton or neutron, averaged across the Universe, is the value that makes the predictions accord with observations. We will be able to cross check this, using a totally different method, in Chapter 8.

Big Bang Nucleosynthesis ended after a few minutes, and laid down the primordial elements that are spread across the Universe today, usually as clouds of atoms but occasionally as stars. Here, we ought to say a word or two about the origin of the other, heavier, elements. These were produced much later on in the history of the Universe, as a result of nuclear processes in the hearts of stars. The first stars were formed out of the primordial hydrogen and helium and, as they burned, they fused the nuclei in their cores to make

---

[8] The theoretical prediction is actually a little high for lithium, when compared to the data. But it is not an easy thing to measure and there is quite a large uncertainty on the physics, so it isn't regarded as too serious a problem.

heavier elements such as carbon and oxygen. Elements as massive as iron can be made this way, but the heavier elements, including gold and silver, were formed in even more spectacular fashion. After a few billion years, massive stars run out of fuel and die. The most massive explode, producing supernovae; the furnaces in which the heaviest elements are forged. They also scatter the newly synthesized elements across space, seeding the eventual formation of planets such as ours. This is why the outer layers of the Sun contain a slightly higher helium content than the oldest stars in the Milky Way: the stars that lived and died before the Sun formed have contaminated it with extra helium, not to mention the oxygen, nitrogen, iron and other elements beyond hydrogen that we see in the solar spectrum. Again, we see consistency in our description of Nature.

Let's now follow the evolution of the Universe forward in time from those first few minutes and see if there is anything else we can deduce that may lead to observable consequences and give us still more confidence in the idea of the Big Bang. The next major landmark occurred when the Universe had cooled sufficiently for the electrons and primordial atomic nuclei to stick together to make atoms. This happens at a temperature of a few thousand degrees, and we understand the process very well indeed from laboratories on Earth. Before the Universe cooled to this temperature, it was largely featureless plasma made up mainly of photons and neutrinos, with a tiny admixture of light nuclei and electrons. At early enough times, when the plasma was too hot, the electrons and nuclei were zipping around too rapidly to stick together and form atoms (they can stick together as a result of their mutual electromagnetic attraction because they have opposite electric charge). But, as the Universe cooled, the particles moved around less energetically and it became increasingly likely that an electron found itself bound in orbit around a nucleus to make an atom. This landmark event – the formation of atomic hydrogen and helium atoms – had a very dramatic impact on the photons in the Universe.

Everything changed when the atoms formed. Before this time, photons were unable to travel very far before hitting an electron or an atomic nucleus. That is because photons scatter strongly off electrically charged particles.[9] But with virtually all the charged particles combining to make electrically neutral atoms, this state of things altered: photons do not scatter so easily off atoms. The Universe therefore made a transition from being opaque to being transparent. Occurring 380,000 years after nucleosynthesis, this is known as the time of recombination.[10] It is a key moment in the history of our Universe and it has left a glorious imprint on the sky.

Figure 6.2 shows three snapshots of the history of the Universe. In the first frame, we trace the paths of photons as they bounce around off charged particles in the hot plasma. In the second frame we have wound the clock forward past the time of recombination: the photons are now travelling in straight lines. These photons carry on until they collide with something, and – because the Universe is pretty empty and filled with an ever-diluting gas of neutral hydrogen and

[9] Precisely how photons interact with charged particles is the stuff of Quantum Electrodynamics (QED), the theory developed by Richard Feynman, Julian Schwinger and Sin-Itiro Tomonaga in the late 1940s. QED is explored in our book *The Quantum Universe*.

[10] As for the case of nucleosynthesis, to calculate this time we need to solve the Friedmann equation. We also need to use simple atomic physics to determine the temperature at which electrons tend to stick to nuclei and make atoms.

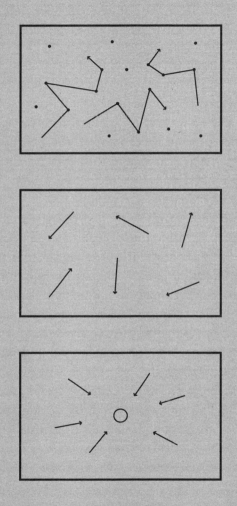

Figure 6.2 The origin of the Cosmic
Microwave Background (CMB) radiation.

helium – most of them keep on travelling until the present day. The last of the three frames shows some of these ancient photons arriving at Earth, now. The possible existence of these photons is a very striking idea. In the words of the Princeton physicists Robert Dicke, Jim Peebles, Peter Roll and David Wilkinson, back in 1964: 'Could the universe have been filled with black-body radiation from [a] possible high-temperature state?'

Dicke and his colleagues made these observations having just heard the news that, away at the Bell Laboratories in New Jersey, Arno Penzias and Robert Wilson had detected an unexpected background of microwaves through a telescope. The fact that the photons from the Cosmic Microwave Background (CMB) should today be observed as microwaves is exactly as expected, because the expansion of space since the time of recombination redshifted the photons to microwave wavelengths. By the mid-1960s, the evidence for what is now known as the CMB had become compelling. The Earth, in other words, is bathed in microwave radiation of precisely the kind anticipated if the Universe had indeed passed through a hot phase before the formation of atoms.

Interestingly enough, the idea that the Universe should today be filled with CMB radiation goes back decades before Penzias and Wilson's Nobel Prize-winning discovery. In 1948, in a tremendous burst of creativity, George Gamow, Ralph Alpher, Robert Herman and Hans Bethe wrote a series of eleven research papers in which they developed the theory of Big Bang Nucleosynthesis.[11] Gamow and colleagues' ideas

[11] For a fascinating but technical review see P. J. E. Peebles, 'Discovery of the Hot Big Bang: What Happened in 1948', *European Physical Journal*, H39 (2014), pp. 205–23.

were ahead of their time and lay largely dormant until the discovery of the CMB in 1964.

The CMB and its properties have been, and continue to be, examined in detail. This oldest light is a treasure trove of information, because it provides Earth-bound astronomers with an opportunity to garner information on what happened shortly after the Big Bang. We'll be delving much further into the treasures of the CMB in Chapter 8, but for now we are going to focus on the broad features of those old, stretched photons from the beginning of time, which we now detect on Earth as microwaves.

Figure 6.3 shows a measurement of the CMB made in the early 1990s by the Cosmic Background Explorer (COBE) satellite. It shows how the microwaves arriving at the Earth are distributed in wavelength. The height of the curve measures the brightness of the microwaves and the shape tells us how this is distributed across different wavelengths. We see that the most commonly occurring wavelength is around 2 mm. What is clear from the graph is that the measured data points are in perfect agreement with the theoretical expectation, which is the smooth solid line.

That smooth line is the result of a calculation originally performed in the late nineteenth century by Max Planck, the brilliant German physicist who also played a central role in the early development of quantum physics. It is the curve corresponding to the spectrum of light emitted by an object cooled to a temperature of 2.73 kelvin, which is just over minus 270 degrees celsius. This amazing agreement between Max Planck's calculation and the COBE measurements is what Dicke and colleagues anticipated when they spoke of 'black-body radiation'. Here, the technical term 'black-body'

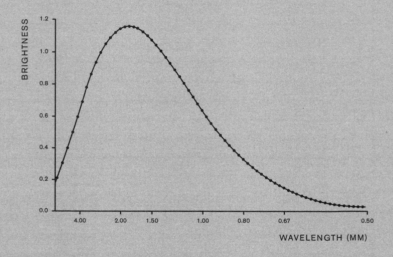

Figure 6.3 The COBE measurement of the
Cosmic Microwave Background radiation, which
shows that the Earth is bathed in microwaves at
a temperature of 2.74 kelvin.

is used to describe how the energy of a gas is shared out between its component particles when all parts of the gas are at the same temperature. Because the CMB photons originated from the primordial plasma, they ought to exhibit a near-perfect black-body spectrum, and this is exactly what was seen by COBE.

The predicted observation of a near-perfect black-body spectrum of microwave radiation is our second, very compelling, piece of evidence in support of the Big Bang.

In a nutshell, then, Big Bang Nucleosynthesis and the existence of the Cosmic Microwave Background provide compelling evidence in support of the idea that the Universe was once, long ago, in a hot, dense state. However, we can dare to go much further because Einstein's equations are also able to provide us with a precise description of the evolution of the Universe, from its earliest moments to the distant future, if we can measure the constituents of the Universe today. In order to do this, we must endeavour to weigh the Universe.

7.
WEIGHING THE UNIVERSE

We now turn to the task of making an inventory of the Universe. We want to know what types of matter exist in it and in what proportions, how much energy is carried by electromagnetic waves, and whether there is anything else out there – anything that is especially hard to detect. This inventory is crucial because Friedmann's equation tells us precisely how these different sorts of material contribute, via their gravitational effects, to the expansion of the Universe. We will be able to compute how the scale factor has changed with time, and how it will change in the future. This will tell us precisely when the Big Bang happened and what the future holds; will the Universe expand for ever or will it eventually collapse? Answering these two momentous questions will in turn allow us to identify which of Robertson's imagined universes we actually live in. Understanding the history of the expansion of the Universe is now our focus – but, as is so often the case in science, we will encounter some surprises along the way.

We described the Friedmann equation in Box 10 (pp. 138–40), and it's worth a look now if you haven't already done so. You don't have to understand the equation in order to understand the rest of this chapter. All you need to appreciate is that the precise way in which the Universe expands is determined by the type and amount of stuff it contains.

From the observation that the Cosmic Microwave Background (CMB) is a gas of photons at a temperature of 2.73

kelvin, we can use some undergraduate-level statistical mechanics to deduce that, today, there are an average of 410 CMB photons in every cubic centimetre of space. In turn, this implies that the total energy carried by these photons is just over 40 millionths of a joule inside each 1 kilometre cube (4×10^{-14} J/m³). This is a tiny energy density by every-day standards: a 10 watt light bulb emits 10 joules of light energy every second.[1] The gravitational influence of this encrgy density is such that it tends to slow down the rate at which the Universe is expanding.

For accounting purposes, we will convert this energy density into mass density, by dividing it by the speed of light squared: doing this gives a mass density equal to 4.5×10^{-31} kg/m³. The meaning of an average mass density is clear for particles that actually do have mass (like protons or electrons): we just need to count up the total mass of particles inside some region of space and then divide it by the volume of that region. Photons are different in that they carry energy but do not have any mass. This peculiarity is a feature of Einstein's Special Theory of Relativity, the details of which are unimportant here: all we need to know is that the energy carried by the photons has a gravitational impact on how the Universe evolves, and that its impact can be quantified by a contribution to the total mass density[2] of the Universe that is today equal to 4.5×10^{-31} kg/m³.

As we saw in Figure 5.4, there are three possibilities for the geometry of the Universe under the assumption that

---

[1] The CMB photon energy density is much larger than that for photons of non-CMB origin, such as those from stars.

[2] Referring back to Box 10 (pp. 138–40), the photon mass density contributes to the quantity denoted by ρ on the right-hand side of the Friedmann equation.

matter and energy are evenly distributed (we called this a homogeneous and isotropic Universe). If the average mass density is precisely equal to a very special value, known as the critical density, the Friedmann equation tells us that space is flat. If the average mass density exceeds the critical density then the Universe curves into a spherical geometry, and if it is smaller than the critical density then the Universe is hyperbolic.[3] If we take the value of the Hubble constant that we obtained from our analysis of spiral galaxies, $H = 70$ km/s/Mpc, then the critical density is 9×10^{-27} kg/m³. There's a nice way to think about this, because a proton weighs in at 1.67×10^{-27} kg: the critical density corresponds to an average of just over 5 protons in every cubic metre of space (there's nothing special about using protons in this comparison: it is like saying that a typical Asian bull elephant has a mass equal to that of around 30 human beings). As it stands, the photons account for far less than the critical density, which means that if there was nothing else in the Universe besides photons we would be living in a Universe with hyperbolic geometry, which would go on expanding for ever. But, of course, the Universe is made of more than just light.

Because we know how much energy in the Universe is carried by the photons, we can immediately deduce how much is carried by neutrinos. This is because in the seconds-old Universe the neutrinos and photons were both bouncing around at the same temperature, together with the rest of the particles in the Universe. The neutrino energy density is

[3] You can see how the density of the Universe controls its geometry by studying Box 10 again; for example, if ρ is bigger than $3/(8\pi G)H^2$ then K must be a positive number, which corresponds to a spherical geometry. You can put the numbers in to determine the numerical value of this critical density and check that it is the value quoted in the text.

slightly less than that due to the photons, for reasons we describe in Box 12 (p. 176). Together, the neutrinos and photons account for a present-day mass density of approximately 7.5×10^{-31} kg/m³.

We noted in the last chapter that the Big Bang Nucleosynthesis calculations agree with the observed data only if there are 1.7 billion photons in the Universe today for every proton or neutron. From our study of the CMB, we have just discovered that there are 410 photons in every cubic centimetre of space. Taken together, these pieces of information mean that, on average, there should be approximately 1 proton or neutron in every four cubic metres of space. Now, protons and neutrons should dominate the mass in the Universe arising from ordinary matter, because they are each around 2000 times heavier than the only other ordinary matter particle; the electron.[4] Suddenly, we find ourselves arriving at a prediction: the matter in the Universe out of which every star, gas cloud and galaxy is made should amount to a grand total of 4×10^{-28} kg/m³. If this is not found to be the case then the Big Bang theory we have been describing will fail.

This is pretty impressive stuff: we are laying claim to knowing precisely how many photons, neutrinos, protons, neutrons and electrons are present on average in every cubic metre of the Universe today. Moreover, we claim to know how those protons and neutrons combined to make hydrogen, deuterium and helium with a smidgeon of lithium. These are the ingredients that have clumped together to make everything we are familiar with. Of course, we would

[4] We need the word 'ordinary' because, as we will shortly see, there's more to the Universe than atoms, electrons, photons and neutrinos.

BOX 12. THE ENERGY DENSITY OF NEUTRINOS   P. 176

As the Universe cooled, neutrinos stopped interacting
with everything else before photons did because the weak
nuclear force acts over much shorter distances than the
electromagnetic force does. So, as the Universe expanded,
it became sufficiently dilute that the neutrinos no longer had
the opportunity to interact. This happened at a temperature
of around 10 billion kelvin, around 10^{-5} seconds after
the Big Bang. Even though they stopped interacting, the
temperature of the neutrinos fell as the Universe expanded,
just as the photon temperature fell. Today, the neutrino
temperature is slightly less than the photon temperature
because, after the neutrinos stopped interacting, the photon
gas got heated up a little bit as a result of a process called
electron-positron annihilation. This is the process by which
an electron encounters a positron (an anti-matter electron)
and both disappear, to be replaced by two photons.[1] In this
way, some extra photons were produced which heated up
the photons a little bit compared to the neutrinos. If you're
wondering how the photons and neutrinos could have two
different temperatures, then bear in mind that this is possible
because the neutrinos have effectively stopped interacting
with anything, which means that there is then no way for
them to share energy with the other particles in the Universe.
The result is that the primordial neutrino temperature today
is slightly less than 2 kelvin. The neutrinos form a Cosmic
Neutrino Background in the sky, at a redshift of around 10^{10},
and it would be fantastic to be able to build a detector to
see it. Unfortunately, that is way beyond current capabilities.

[1] You might recall from Chapter 2 that electron-positron
annihilation was responsible for 1.02 MeV of energy in the
fusion chain that leads to helium production inside the Sun.

love to be able to corroborate these numbers using some entirely different piece of science, and it is one of the highlights of modern cosmology that – as we'll see in the next chapter – we are able to do so. For now, we must press on, because there there is more to the Universe than just light and ordinary matter.

The earliest strong evidence that there is some other form of matter beyond that visible through telescopes was presented way back in 1933, in a paper written by the Swiss astronomer Fritz Zwicky. He noticed that the galaxies in the Coma galaxy cluster[5] were moving much faster than expected. The galaxies that make up a cluster are in orbit around each other due to their mutual gravitational attraction, just as the Earth orbits the Sun, except that the distances involved are vastly greater: typically the galaxies are a few million light years apart. Using Newton's laws, we can estimate the total mass of the cluster if we know how fast each galaxy is moving. This is rather like the example we encountered in the previous chapter when we discussed the Tully-Fisher relation for stars orbiting around a galaxy. It's also similar to the way we inferred the mass of the Sun using Newton's laws and observations of the planetary orbits in Chapter 3. If we replace the word 'star' by 'galaxy' and 'galaxy' by 'cluster' in Box 11 (pp. 148–9), we can deduce the speed of an average galaxy in a cluster by observing the Doppler shift of the light it emits. Once we have the speed of the galaxy, we can estimate the mass of the cluster. Zwicky discovered that the mass of the Coma cluster is far bigger than

[5] The Milky Way is a member of a different galaxy cluster, known as the Local Group. It contains around 50 small galaxies plus 2 big spiral galaxies (the Milky Way and Andromeda).

can be accounted for by its visible contents. The light emitted by the cluster is roughly equivalent to that emitted by 30 trillion suns, while the inferred mass is equivalent to that of 4500 trillion suns. That is quite a difference.

Astronomers speak of the 'mass-to-light ratio', which is the total mass divided by the total light output. In the case of the Coma cluster, this is approximately 4500/30 = 150. Our Sun has a mass-to-light ratio of 1 in these units.[6] Such a large mass-to-light ratio immediately tells us that most of the mass of the Coma cluster is not located in stuff that emits light. It is natural to suppose that this might be in the form of dark objects like dead stars or black holes, but we will see in a moment that this cannot be the whole story. Whatever the case, there is a lot of dark material to account for – and this was a surprise to Zwicky. What's more, we now know that there is nothing special about the Coma cluster. The largest clusters have mass-to-light ratios of around 250.

The mass-to-light ratio is a particularly valuable number. If we know its value averaged across the entire Universe we can use it in conjunction with the measurement of the average luminosity of space to infer the average mass density of space. The average luminosity of space corresponds to 130 million suns per cubic megaparsec. So, if the mass-to-light ratio for the Universe is 250, it follows that the matter density is equivalent to 250×130 million suns inside each 1 Mpc cube. Given that our Sun has a mass of 2×10^{30} kg, this gives an average matter density of 22×10^{-28} kg/m^3. You may reasonably object that the mass-to-light ratio of a cluster of galaxies is not the same thing as the mass-to-light ratio of the

[6] This corresponds to 5100 kg/watt.

Universe at large, which we cannot measure directly. This is true: indeed, for our solar system the mass-to-light ratio is very close to 1, because the Sun carries most of the mass and gives off most of the light. Conversely, in the vicinity of a black hole or dead star, the mass to light ratio will be very large because there is a lot of mass but not much light. However, the larger the region of space over which we average, the more we expect it to approximate the behaviour of the Universe at large, so by looking at the very largest galaxy clusters we expect to obtain a decent estimate of the average mass-to-light ratio of the Universe.

A mass density of 22×10^{-28} kg/m^3 is more than 5 times larger than the number we determined for the mass density of ordinary matter, which implies that the material content of the Universe is predominantly composed of something other than the products of Big Bang Nucleosynthesis. This is important because it removes the possibility that the unseen matter might arise from dead stars or black holes or other non-luminous stuff made from atoms, because the mass density of ordinary matter that we estimated from the Big Bang Nucleosynthesis was not just restricted to the matter that we can see. We call this new stuff dark matter.

This is, of course, a rather bold conclusion to draw. We seem to have invented a new form of matter simply because our calculations and observations don't match. This would be a valid criticism if it weren't for the fact that there are many other ways of measuring the amount of matter contained within galaxies and galaxy clusters – and they all give the same result. During the 1970s, a series of observations pioneered by the American astronomer Vera Rubin revealed that, for very many galaxies, stars and gas are orbiting much faster than would be expected if the only mass in the galaxy were luminous. The physics is the same as for galaxy clusters – except here it is stars, rather than entire galaxies, that appear to be moving too fast. Analysis of Rubin's 'galaxy rotation curves' indicates that as much as 90% of the mass in a galaxy is not luminous, and until quite recently this was the principal evidence for dark matter.

The most dramatic evidence for the existence of dark matter comes from observations of the aftermath of a violent cosmic collision, as two clusters of galaxies ploughed through each other at several million miles per hour. Plate 18 shows several images of what is now known as the 'Bullet Cluster' superimposed on top of each other. The background is a composite image taken in visible light by the 6.5 metre Magellan telescopes in Chile and the Hubble Space Telescope. The red clumps are an image of the same system taken by the Chandra X-ray observatory and they show where the hot gas resides (by detecting the X-rays they emit). In the Bullet Cluster, this gas is glowing at a temperature of around 100 million degrees. The blue clumps show the location of the majority of the mass in the cluster, inferred using a technique known as gravitational lensing, an ingenious application of General Relativity. Space and time are distorted by the presence of the matter in the Bullet Cluster. This means that the image of anything behind the Bullet Cluster, as viewed from Earth, will be distorted because the light has to pass through the distorted spacetime on its way to us. Plate 19 shows a more dramatic example of gravitational lensing. The galaxy cluster Abell 2218 is distorting the images of more distant galaxies lying behind it, which are visible as smeared-out arcs of light. The effect is rather like viewing something through the bottom of a wine glass. Just as we can infer the shape of a wine glass from the way it distorts an image, so we can infer the shape of spacetime by the way it distorts an image. General Relativity, then, allows us to determine the distribution of matter in the intervening galaxy cluster, because it tells us what distribution is necessary to distort spacetime in the required way. Incidentally, the masses of galaxy clusters can also be determined using gravitational lensing, and the results agree very well with the masses inferred using the orbital speeds of their constituent galaxies in the manner first used by Zwicky in the 1930s. As ever, it is good to be able to measure the same thing in different ways.

It is immediately obvious from Plate 18 that the majority of the mass in the Bullet Cluster is not in the vicinity of the hot gas. This is inexplicable without invoking dark matter, because the bulk of the mass *that we can see* is located in the vicinity of the hot gas – meaning that, unless dark matter is present, the blue and red regions should lie on top of each other.[7] Clearly, the rightmost cluster of galaxies has ploughed its way from left to right: you can see the shock-wave of X-rays lying in its wake. During the collision, the hot gas of charged particles in each cluster was slowed down, which is expected because the hot gas is composed of electrically charged particles that interact strongly and

---

[7] You may have spotted that there are not many galaxies in the region of the hot gas (the galaxies are in the blue regions, in fact). Do not be confused, though: it is very well known to astronomers that the bulk of the ordinary mass (i.e. excluding dark matter) in the Universe is in gas and not in stars. The brightness of the X-rays allows astronomers to determine how much mass resides in the gas (more gas means more X-rays), and this is considerably more than resides in the galaxies of stars. The galaxies in Plate 18 follow the dark matter because, unlike the gas, they are unlikely to interact with each other much. The galaxies are not the important parts of the Bullet Cluster photo.

scatter off each other. By contrast, most of the mass in the clusters was evidently much less affected by the collision. Indeed it hardly seems to have felt the collision at all; it has continued on its journey through space, passing straight through another galaxy cluster at over a million miles an hour. The collision has caused the majority of the ordinary matter in the clusters, contained in the hot gas, to become separated from the majority of the mass, which exists in the form of weakly interacting dark matter.

Today, the existence of dark matter is very well established, although we certainly do not know what it actually is. The simplest idea would be to suppose that there is a new type of particle, which hasn't been observed on Earth yet. This is a very reasonable possibility: it would be verging on the arrogant to suppose that the only particles that exist are those we have already observed. As we have seen, the dark matter only appears to interact appreciably via gravity, which is why it is dark. It could conceivably have some weak, non-gravitational interaction with other particles, which might make it possible to eventually produce and/or detect it in particle physics laboratories on Earth. That possibility has been considered very seriously, not least because some of the more popular ideas in particle physics do predict the existence of just such a dark matter particle, and those ideas do not draw on any motivation from astronomy and cosmology. As we will see in the next chapter, dark matter can also be inferred from the way that galaxies clump together across the entire Universe and from an analysis of the fine details of the Cosmic Microwave Background.

Now is a good time to pause for breath and recap where we are in our quest to pin down the material contents of the

Universe. We have worked out that, at the present time:

(i) photons and neutrinos give rise to an average mass density of 7.5×10^{-31} kg/m³;

(ii) there are approximately 4×10^{-28} kg/m³ of ordinary matter, which is made up of atomic nuclei and electrons;

(iii) the sum total of the dark matter and the ordinary matter averages to 22×10^{-28} kg/m³, which means that the dark matter alone contributes around 18×10^{-28} kg/m³.

Because the sum total of these mass densities is about 30% of the critical density (9×10^{-27} kg/m³), the possibility that we live in a hyperbolic Universe is still looming. However, there is something very curious to note: the critical density corresponds to around 5 protons per cubic metre, while the sum total of the dark and ordinary matter checks in around 1.5 protons per cubic metre. Why are these numbers so similar? Without any prior bias, we would have had no trouble imagining that they might be wildly different: a trillion protons per cubic metre, for example, or maybe one proton every megaparsec. The fact that the observed density is so close to the critical density smacks of coincidence – or perhaps it is a clue. Whatever the case, it certainly means that our Universe is not so far from being flat.

This apparent coincidence is all the more surprising because, according to the Friedmann equation, the difference between the matter density and the critical density

[8] You can't spot this just by looking at the equation; you need to solve it. Roughly speaking, as time passes we become increasingly aware of any curvature to space (i.e. the density increasingly deviates from the critical density) – so if we cannot discern any curvature now then we most certainly couldn't discern any in the past (i.e. the density was closer to critical in the past).

ought to get bigger as time passes.[8] This means that if the density is close to critical now, which it is, then it must have been even closer to critical in the past. This is an example of what physicists call a 'fine tuning' problem. It is as if the matter density in the Universe was adjusted very precisely at the Big Bang in order to give rise to the Universe we see today. To put it another way, for us to find ourselves in a nearly flat Universe now, space must have been 'unfeasibly close to flat' at the Big Bang. Cosmologists refer to this as 'the flatness problem'.

Many cosmologists would much rather contemplate a perfectly flat Universe in which the density is equal to the critical density than a nearly flat one. At first sight, this might sound like a prejudice; why would a scientist be happy with a Universe of precisely the critical density rather than 30% of it? The explanation for the apparent prejudice is that there is a very simple reason why the Universe might appear to be flat. Specifically, even a curved universe would appear flat if it is sufficiently big.

The link between flatness and the size of the Universe is easy to grasp. A person restricted to roam over a small portion of the Earth's surface could easily be fooled into thinking that the Earth is flat simply because it is big. Likewise, if we are doing our astronomy in a small portion of a vastly bigger Universe, then we could easily be drawn to conclude that it is flat. You can see this idea at play in the Friedmann equation (take a look at Box 10 again): making R larger has the effect of making the term that depends on the geometry of space smaller, and one can imagine it being so large as to make its effect irrelevant. In that case, the Universe would appear flat even though it isn't flat on the largest scales. The question of why the density of the Universe should be close to the critical density then gets replaced with the question of why the Universe is so big. As we've suggested in Chapter 1, a huge Universe is the natural consequence of the theory of inflation. Perhaps you now see the reason why a cosmologist might prefer to suppose that we live in a very flat Universe, whose mass density is very close to the critical density, rather than one where it is 30% of the critical density. In the super-flat case, the task is 'simply' to find a theory (like inflation) that delivers a Universe that is vastly bigger than the observable Universe. This is rather easier to contemplate than trying to find a theory that delivers a 'just so' Universe, where the radius R just happens to be roughly the same size as the observable Universe. Obviously, the trouble with supposing that we live in a very flat Universe instead of a nearly flat one is that we need to account for the missing 70% of the mass density. Remarkably, the theoretical prejudice for a very flat Universe appears to be correct, although the evidence to prove it comes from a quite unexpected direction: towards the end of the twentieth century, Einstein's cosmological constant rose phoenix-like from the ashes.

We met the cosmological constant back in Chapter 5, when we observed that Einstein introduced it into his equations of General Relativity in an unsuccessful attempt to construct a viable model of a non-expanding universe. After Einstein dismissed it as his greatest blunder, the possibility that the Universe might today be endowed with a cosmological constant did not feature high on the priorities of most cosmologists – not until the 1990s, that is.

The Friedmann equation allows us to calculate the age of the Universe (once we know the average mass densities of the different types of matter and the present-day Hubble constant). The trouble is that a density of matter equal to only 30% of the critical density and a Hubble constant of 70 km/s/Mpc suggests that the Universe is younger than some of its contents, which is obviously a bit of a disaster. The problem comes from the dating of clusters of stars known as globular clusters.

Astronomers love globular clusters because they provide some of the most spectacular sights in the night sky that can be viewed through a small telescope. They are great balls of stars, and are numerous in the Milky Way. The left-hand image in Plate 20 is a photograph of Messier 3 in the northern constellation of Canes Venatici. This picture was taken by a friend of ours, Bill Chamberlain, who recently took up astronomy as a hobby at the tender age of 70. One of the most spectacular and well-studied globular clusters is Omega Centauri, the largest globular cluster in the Milky Way. It is only 15,800 light years away from Earth, and contains ten million stars, which makes it a bright and beautiful object. A quite stunning photograph of Omega Centauri, taken by the European Southern Observatory's La Silla Observatory, is also shown in Plate 20. Astronomers can determine the age of clusters like M3 by studying the variation in colour and brightness of the stars that make up the cluster. In the case of M3, they have figured out that it is around 11.4 billion years old.

To study systems of stars, astronomers arrange the stars into a diagram according to their colour (which is directly related to their surface temperature) and brightness. These

---

[9] The HR diagram is named after Ejnar Herzsprung and Henry Norris Russell, who first presented stellar data in this way, independently from each other, in 1911.

are known as Herzsprung-Russell (HR) diagrams:[9] Plate 21 shows the HR diagrams for Omega Centauri (right) and the Pleiades (left). These make for very pretty diagrams because the stars are obviously not scattered about at random. In the case of the Pleiades, the majority of the stars lie on the same, sweeping curve, with the hot, blue stars being the brightest and the cooler, red stars being the dimmest. This curve is known as the 'main sequence'. The main sequence is less prominent in the case of Omega Centauri. Instead, the pattern of stars curves back on itself, indicating that Omega Centauri contains a significant population of bright, red stars. These stars are known as red giants.

The basic physics of the HR diagram is simple to grasp. Main sequence stars like our Sun are busy fusing hydrogen into helium in their cores. The energy released in these fusion reactions creates the pressure that resists the inward pull of gravity, allowing the star to exist in a stable state as long as it has fuel to burn. More massive stars have to generate more heat energy to create the higher pressure needed to resist the inward pull of gravity; they are therefore hotter and bluer, at the expense of having to burn their fuel more quickly. In contrast, less massive stars have to generate less energy, and are therefore dimmer, redder and burn their fuel more slowly. As stars run out of hydrogen fuel, they contract and their cores heat up, allowing them to initiate helium burning. This is how some of the heavier elements, including oxygen and carbon, are made. Their super-heated cores cause the outer layers of the star to expand and cool, resulting in the cool but bright stars – red giants – which can be seen towards the top right of the HR diagrams. The story of stellar evolution, then, is laid out on an HR diagram. For us, the key observation is that when a star runs out of hydrogen fuel it will move off the main sequence and onto the red giant branch, which is much more prominent in the case of Omega Centauri than it is in the Pleiades.

Bright blue-white stars will move off the main sequence first, followed by yellow stars like our Sun, leaving only the very long-lived dim red dwarf stars behind: these have enough fuel to carry on shining for many times the current age of the Universe. In a relatively short time, the red giants will exhaust their fusion fuel completely, at which point their cores will collapse into either a white dwarf or, for the most massive stars, a neutron star or a black hole. White dwarfs are dim, blue-white stellar remnants, and sit in the lower left of the HR diagram. Perhaps you can now see how an HR diagram might allow astronomers to date a star cluster. Older clusters will contain a larger number of red giants and white dwarfs, and fewer stars on the main sequence; whereas, for very young clusters, all the stars will still be on the main sequence and there will be no red giants.

With only this information, it is clear that the Pleiades is significantly younger than Omega Centauri because it contains no red giants, and a large population of young, bright blue stars; meanwhile, red dwarves, red giants and a smattering of white dwarves are common in Omega Centauri. Astronomers can do much better than simply determining the relative ages of clusters, because they understand stellar evolution. In Chapter 2, we worked out the age of the Sun using our knowledge of nuclear physics, and we can do the same with the stars in clusters. If you look at the HR diagram for Omega Centauri, for example, you see that there are no yellow Sun-like stars left on the main sequence.

Because we know that yellow main-sequence stars like our Sun have a lifetime of around 10 billion years, this must mean that Omega Centauri is older than 10 billion years; it is, in fact, 11.5 billion years old.

This age is interesting because it is marginally older than the age of a universe that is currently expanding at 70 km/s/Mpc and contains only matter with a density equal to 30% of the critical density – using the Friedmann equation, such a universe would be 11.3 billion years old. The problem is even more serious because there are other globular clusters that have ages in excess of 12 billion years. At a stroke, the globular cluster age problem can be solved by supposing that the Universe is endowed with a cosmological constant, whose size corresponds to a mass density equal to 70% of the critical mass density. This has the twin advantages of increasing the predicted age of the Universe to closer to 14 billion years and increasing the total mass density to 100% – as befits a flat Universe. By the mid-1990s other data, such as that coming from the theory of galaxy formation that we will encounter in the next chapter, was also encouraging cosmologists to take seriously the possibility that there might be a cosmological constant after all. However, for many cosmologists the balance finally tipped in favour of the existence of a cosmological constant in 1998.

In that year, two teams of astronomers presented their results on the redshift-distance relationship for Type 1A supernovae. In other words, they made a Hubble plot just like we did to measure the current expansion rate of the Universe in the last chapter, but instead of using spiral galaxies the astronomers used Type 1A supernovae. Recall that these exploding stars are especially valuable to astronomers because they all explode in the same way, which means we can measure how far away they are using their observed brightness. Because supernovae can be seen at very large distances, they can be used to observe how the Universe has been expanding over the past few billion years, which is far better than we managed using spiral galaxies.

The Supernova Cosmology Project and the High-Z Supernova Search Team published their observations after several years of collecting data on a few tens of supernovae. They found, independently, that most distant supernovae were significantly dimmer than expected, which is to say that they were further away than they should be in a universe dominated by matter alone. The best fit to the data in fact suggested that the Universe is not decelerating but accelerating in its expansion. In a universe containing only ordinary and dark matter, the gravitational attraction of the matter tends to cause the expansion rate to gradually slow down, so deceleration would be the norm – an accelerated rate of expansion is the hallmark of a cosmological constant. Without doubt, the most remarkable aspect of the supernovae measurements is that they can be explained if the mass density associated with a cosmological constant is equal to 70% of the critical density. In other words, the Universe really does appear to be flat – the theoretical hunch was right.

The cosmological constant now often goes by the name of dark energy. The more mystical-sounding name is suggestive of the possibility that there might be a deeper explanation for what causes the accelerated expansion of the Universe – perhaps something exists that mimics the effect of a cosmological constant. Many ideas abound, but so far there are no compelling explanations for what dark energy is – it could be nothing more than a cosmological constant. One idea is that empty space itself is a source of energy (called vacuum energy) – it is an idea that is very familiar to particle physicists, and we will meet it again in the next chapter. But there is a serious problem with the notion of vacuum energy: the particle physics calculations tend to predict a cosmological constant that is vastly bigger than observed. For this reason, particle physicists tended to suppose that the cosmological constant is actually equal to zero and that their understanding of vacuum energy is flawed. The preference for a value of zero comes about because it is easier for a theorist to imagine that some presently unknown physics magically cancels away all of the large numbers leaving precisely nothing than it is to imagine some presently unknown physics that even more magically cancels away *almost* all of the large numbers leaving *almost* nothing. Admittedly that isn't the most compelling logic – but the inability of theorists to make sense of the cosmological constant meant that they were very wary of claims that it might actually be present in Nature. The fact that the data do seem to require a non-zero value of the cosmological constant forced them to face their demons. The theoretical prejudice against a cosmological constant was initially such that the leader of the High-Z Team, Brian Schmidt, who shared the 2011 Nobel Prize for the discovery, has said that he thought the publication of the result would be the end of his career, because he felt it must be wrong.

Here, then, is the present-day inventory of the Universe that we've established in this chapter: the mass density is (in percentages of the critical density) divided into 5% matter, 25% dark matter and 70% dark energy. We can now settle the question as to which one of Robertson's universes we inhabit. We live in universe type (i): our Universe is of a finite age, and it will expand for ever into the future.

Now that we know how the mass and energy in the Universe are shared between its different components we can, using Friedmann's equation, compute the rate at which the Universe expands, for all times after the Big Bang. Figure 7.1 shows how the scale factor[10] varies with time for four conceivable universes, with varying amounts of matter and dark energy: our Universe corresponds to the black curve.

The present day corresponds to a scale factor of 1, and we can therefore read off that our Universe is just less than 14 billion years old. To be precise – and we need to be for what is coming in the next chapter – we mean that this length of time has passed since the scale factor was very small. We don't actually believe that the Friedmann equation is correct when the scale factor becomes too small, but we do trust it to the point where the temperature of the Universe corresponds to the energies being probed at the highest energy particle collider on Earth, the Large Hadron Collider. These energies correspond to a temperature of around 10^{16} degrees celsius, which occurred at a tiny scale factor of 10^{-16}. This corresponds to a time when the current visible Universe was about $10^{-16} \times$ 10 billion light years across (= 10 million km). Although we appear to be backing away from claiming an understanding of the very origins of the Universe, we are still making the audacious claim that we can now describe its evolution starting out from a time when all the matter necessary to make all

---

[10] Recall this is the number that says how much space is stretched or shrunk relative to its size today.

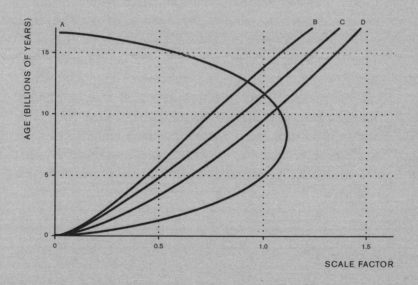

Figure 7.1 The variation of the scale factor with time. Curve
B is our Universe, i.e. 30% matter and 70% dark energy.
Curve C is the same but with no dark energy, and D is for a
universe containing 100% matter. A is for a universe with no
dark energy and a matter density 10 times the critical density
– as can be seen, it is a universe that ends with a Big Crunch.

the hundreds of billions of galaxies in the observable Universe would have been contained in a sphere that would sit comfortably inside the Earth's orbit around the Sun. But we are being too modest – we can do better than that.

Oh – but before we do, we almost forgot about weighing the Universe. While a cosmologist would be content with saying that the Universe is at the critical density, we can do the calculation ourselves. The Friedmann equation allows us to ascertain that the observable Universe is contained in a sphere of radius 47 billion light years, which is bigger than 14 billion light years because of the expansion of space.[11] That means the volume of the observable Universe is just under 4×10^{80} m^3. We have ascertained that the average mass density is 9×10^{-27} kg/m^3 (if we include the contribution from dark energy), and so the total mass of the observable Universe is just over 3×10^{54} kilograms.[12] Numbers this size are very difficult to picture – if it helps, we might say that the observable Universe weighs as much as 5×10^{54} pints of bitter.

[11] The fact that 47 billion light years is bigger than the time since the Big Bang multiplied by the speed of light (= 14 billion light years) is because the 47 billion years refers to how far away the edge of the observable Universe is *now*. In the past, this distance was smaller owing to the fact that space expands.

[12] We worked out the volume of the observable Universe using the formula $4/3 \pi r^3$ for the volume of a sphere of radius r, and we worked out the total mass by multiplying this volume by the mass density.

8.
WHAT HAPPENED BEFORE THE BIG BANG?

The Universe is not a homogeneous mulch of matter; it is full of intriguing structure. Stars are gravitationally bound into a beautiful variety of galaxies, and the galaxies are woven into filamentary networks spanning many millions of light years (Plate 22). What we want to do now is to explore how structures like these emerged out of the primordial plasma, the products of the Big Bang – this will lead us to focus on a time when the entire visible Universe was compressed into a region of space much smaller than the nucleus of an atom.

As we have seen, for most of the first 380,000 years after the Big Bang, the Universe was filled with an almost featureless hot plasma composed of electrons, atomic nuclei (mainly protons) and photons.[1] We know the plasma was nearly featureless, because the radiation we detect from this time – the Cosmic Microwave Background (CMB) – is almost perfectly uniform in all directions. It could not have been absolutely featureless, however, otherwise it would have remained so for all time: galaxies would not have been formed. In some places the Universe must have been ever so slightly denser than in other regions, and a closer inspection of the CMB microwaves does indeed reveal that they are not exactly the

[1] Dark matter and neutrinos were also present at this time, but they did not interact much with the particles in the plasma.

same across the sky. The European Space Agency's Planck satellite (see Plate 23) marks the pinnacle in our observations of the Cosmic Microwave Background; Plate 25 shows the magnificent photograph it has taken of the light coming from the time of recombination.

Using computer simulations, we can go ahead and track the evolution of the Universe starting from the time of recombination all the way through to the present day. During this period of evolution, the dark matter in the Universe took centre stage. Those regions where the dark matter was of slightly higher than average density tended to attract more matter from their surroundings. Throughout the Universe's history, this gravitational tendency for dark matter to clump has competed with the expansion of space, which has the opposite effect of diluting matter. Initially, during the first 50,000 years after the Big Bang, the expansion won out because space was expanding too quickly for the matter to clump: after this time, those little lumps of dark matter started to grow, gradually pulling in the ordinary matter as well. In this way, large clouds of atomic nuclei formed that, in the subsequent few billion years, collapsed to make galaxies of stars. This evolution of the Universe, from an almost-smooth distribution of matter into a lumpy one, is an unavoidable consequence of gravity.

Plate 24 shows the results of two such simulations. The image on the left, provided by our colleague Scott Kay, shows how the dark matter should be distributed today across a large portion of a Universe like ours (i.e. with the same amounts of dark matter, dark energy and ordinary matter). The image on the right is a simulation performed as part of the Eagle Project, which involves cosmologists from across

Europe. This image shows how the ordinary matter (mainly hydrogen gas) is spread across a much smaller portion of the Universe; it's a zoomed-in version of the left-hand image, in which the wispy nature of the matter distribution is more evident. Astonishingly, the Eagle computer simulation is able to describe the formation of realistic-looking individual galaxies.

We can compare these simulations to the map of the galaxies in the real Universe from the Sloan Digital Sky Survey, shown in Plate 22. The same wispy distribution of matter is evident. Notice also how the ordinary matter (the stuff of galaxies) tends to clump in networks that resemble those in the dark-matter distribution. As we noted above, this is because the dark matter is more abundant and its gravity tends to attract the ordinary matter. The agreement between simulation and observation looks good to the naked eye, but it also stands up to more rigorous mathematical analysis, meaning that we can be confident in our use of these simulations to better understand the real Universe.

The simulations reveal that the dark matter interacts primarily via gravity (as we suspected), but also that it should be 'cold': cosmologist's jargon meaning that it is composed of particles that are not zipping around at high speeds. Simulations in which the dark matter is 'hot' (as would be the case if the dark matter was composed of fast-moving neutrinos) do not look anything like Plate 22; they even fail to produce galaxies. Through simulations like those of the Eagle Project, we have a good understanding of how the lumpiness in the CMB relates to the distribution of the galaxies in the Universe today. Our next challenge is to understand where this initial lumpiness came from.

In doing so, perhaps unsurprisingly, we are going to return to the theory of inflation that we encountered briefly in the last chapter and in Chapter 1. First, though, we should stress that the intellectual origins of the theory of inflation had nothing to do with cosmologists' attempts to describe the origin of structure in the Universe. That's to say, we're not working backwards from the Universe we see today and inventing a theory to explain the structures within it. Rather, we're using a pre-existing theory that was constructed for entirely different reasons.

Part of the original motivation for the theory of inflation was that it should be able to solve two puzzles that appear at first sight to be completely unrelated: the 'flatness' and 'horizon' problems. We met the flatness problem in the last chapter, and found that it may not be a problem at all, if the portion of the Universe we are currently able to explore proves to be just a tiny (and relatively flat) part of a much bigger (and curved) Universe. Inflation delivers this huge Universe by causing space to expand at a phenomenal rate – we will explain how it did so shortly. Regardless of the details, we know that the period of inflation must have happened before the time when the first atomic nuclei formed, which you will recall was when the Universe was at a temperature of around a billion degrees. If inflation happened during or after nucleosynthesis, the nuclei would have become diluted and this would spoil the good agreement between the theoretical predictions and observations of their abundances that we discussed in Chapter 6.

The horizon problem can be appreciated by thinking about the uniformity of the CMB – the near-perfect black body spectrum we described in Chapter 6. Any gas of interacting particles will tend to share out the available energy between the particles as a result of collisions. Faster particles will lose energy whereas slower ones will gain energy. After a period of time, the gas will settle into an unchanging state, known to physicists as a state of thermal equilibrium. For a gas in equilibrium, the temperature is a measure of the average kinetic energy of the gas particles. In the time before the CMB photons started on their straight-line journeys across space, the Universe was still a hot plasma and the photons were a part of it, jostling the electrons and protons to share out the available energy. The temperature recorded by the Planck satellite is a direct measure of the temperature of the hot gas in those final moments before the photons took flight.

So far so good. Now let's uncover the horizon problem, which is illustrated in Figure 8.1. The CMB photons arriving today at Earth all started out from points lying on the surface of a sphere, whose radius is the distance light can travel in a little under 14 billion years. This is the grey circle in the figure and is referred to by cosmologists as the 'surface of last scattering'. Now, consider two photons arriving at the Earth from opposite points in the sky. These two photons were released from the hot plasma at points A and B in the figure. The Planck data records the photons coming from A and B and informs us that these regions in the plasma were at the same temperature, to one part in a hundred thousand. The question is, how did A and B get to be at so nearly the same temperature when they are so far away from each other and at first sight could never have been in contact with each other? This is the horizon problem, and it is worth describing in a little more detail.

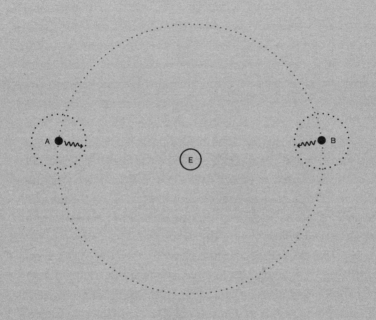

Figure 8.1 Illustrating the horizon problem.

If, before the time of recombination, the expansion of the Universe was always governed by the ordinary and dark matter it contains[2] then, according to the Friedmann equation, the following problem arises. Think about the point in the primordial plasma labelled A. In the 380,000 years that the Universe had been expanding before recombination, this part of the plasma could only ever have influenced other parts of the plasma that lie within a sphere whose radius is the distance that light could travel in those 380,000 years. The small circle around point A shows the size of A's possible sphere of influence at the time of recombination. The same is true for point B. The particles inside each of these spheres could conceivably have had sufficient time to jostle their neighbours, share out their energy and reach a common temperature by the time the CMB photons were released. The problem is that the two spheres of influence are not even close to overlapping, which means that the gas near to A never jostled with the gas near to B. This makes us wonder what caused them to be at the same temperature.

One explanation might be that 'whatever created the Universe in the first place did it in such a way that, at the moment of the Big Bang, everything was created with near perfect uniformity in temperature.' That's fine as far as it goes, but it would be nice to have some kind of explanation beyond this. Inflation provides such an explanation by supposing that, at some time prior to recombination, the Universe underwent a freakish period of very rapid expansion, such that the spheres of influence of A and B were much bigger than indicated in the figure. With a fast enough expansion, they would be big

---

[2] The cosmological constant played a very minor role early on in the Universe's expansion.

enough to overlap. The horizon problem is a little bit like the case of two aliens who know each other despite appearing to be too far away to have ever met. The puzzle is solved once we know that the space between the aliens underwent a rapid expansion sometime in the past, so that the aliens were once neighbours.

Of course it is easy enough to say that a rapid period of expansion[3] early on sounds like a great idea, but 'sounding like a great idea' doesn't count for much. We must check that inflation is a viable theory in detail as well as in a broad, hand-waving sweep. To do this, we need a mechanism for inflation and some predictions to compare with data. One way to arrange for a rapid burst of accelerated expansion is to add a sufficiently large cosmological constant into Einstein's equations. We already know that a cosmological constant causes the expansion of the Universe to accelerate, and the bigger its value, the bigger the acceleration (if you know a little maths, you can see this by solving the Friedmann equation in Box 10 (pp. 138–9) for the scale factor, with only a cosmological constant and neglecting all other terms). As we saw in the last chapter, the evidence even points to there being a cosmological constant today. The trouble is that the present-day cosmological constant is way too small to produce the dramatic expansion needed to solve the horizon and flatness problems. Fortunately, there is a way

---

[3] The horizon and flatness problems are both 'just solved' if the Universe grew during its period of inflation by at least as much as it has grown since the inflation ended. This isn't obviously the case and to show it would be too much of a detour for this book. It means the Universe inflated *at the very least* by a factor of around 500 million, which is the amount the Universe has expanded from the time of nucleosynthesis until today. In practice, most theorists expect that it inflated for vastly more than this – a factor of 10^{26} is commonly regarded as the minimum amount of inflation.

forward: General Relativity can be combined with a mainstream idea in particle physics to produce something that is exactly like a cosmological constant. The idea from particle physics is that of a 'scalar field'.

Fields are familiar things to us all. You can map out the magnetic field produced by the Earth using a compass, and mobile phones work by transmitting and receiving waves in the electromagnetic field. The 'scalar field' that we shall consider is very similar to the electromagnetic field, the important difference being that it can pervade the whole of space as a kind of backdrop to everything else that happens; picture the scalar field as a still ocean filling space. According to General Relativity, if an ocean-like scalar field was present at some time in the history of the Universe, the energy stored within it could act like a cosmological constant and make the Universe expand very much more rapidly than would otherwise be the case. In this 'particle physics' way of thinking, a period of inflation is a very natural thing, because it is what tends to happen if scalar fields exist in Nature. In Box 13 (pp. 204–8), we explore the idea of an all-pervasive scalar field in more detail; we explain that fundamental fields also predict the existence of corresponding particles, and we talk about a scalar field that we know to exist because we discovered it at the Large Hadron Collider: the Higgs field.

The conjectured scalar field that caused the young Universe to accelerate so phenomenally quickly is known as the 'inflaton' field. As far as we can tell, the inflaton doesn't play much of a role in our Universe today. There is no trace of any inflaton particle production at the Large Hadron Collider, for example. Given that scalar fields can cause accelerated

BOX 13. FUNDAMENTAL FIELDS AND THE HIGGS BOSON   P. 204

When physicists think of a field, they imagine some quanti-
fiable thing that is spread out across space. A very simple
example of a scalar field is the temperature field in a room. At
each point in the room we can associate a temperature, and
so can represent the field by a list of numbers, one for each
point in space. If you want to know the temperature somewhere then you just need to look up the correct number from
the list. Obviously this is dull. The utility of the field concept
comes when we want to do more sophisticated things, like
track how the temperature varies with time. If there is a
source of heat in the room, then the air temperature would
change with time in a complicated way: the air close to the
heat source would become warmer and rise upwards by con-
vection and this would disperse through the room, and so on.
If we had the right equations, we could attempt to compute
how the temperature field throughout the room would vary
in the presence of the heat source. In this case, the field is a
concrete mathematical representation of something very real,
and it can be manipulated using mathematics to allow us to
compute the temperature in the room at any place and at any
particular time. This is a good illustration of how physicists
use mathematics: they represent real physical things by
abstract mathematical entities (like fields), and then they process those entities using mathematical operations in order to
answer concrete questions about the real world.

The temperature field is a scalar field because we can
specify it by giving just one number at each point in space.
The electromagnetic field is not a scalar field; it is a vector
field. Vector fields are specified by giving a number and a
direction at each point in space. A simple example is the flow
of the air in a room. At each point we should say how fast the
air is flowing and also in which direction it is heading. Taken
together, these two pieces of information (stated for each
point in the room) correspond to a complete representation of
the air-flow field. The directional piece of the electromagnetic
vector field is referred to as its polarization – exploited, for
instance, in the design of polaroid sunglasses, which selec-
tively block out electromagnetic light waves of a particular
polarization. Roughly speaking, the polarization of a light wave

BOX 13. FUNDAMENTAL FIELDS AND THE HIGGS BOSON   P. 205

tells us the direction in which the wave is waving. Think of shaking waves onto a rope: the direction in which your hand is shaking determines the polarization of the wave. Where the temperature and air-flow fields differ from the electromagnetic, Higgs or inflaton fields is that in the former it is very clear that the fields are representing something more fundamental: they are tracking features of the air in the room. In the case of the electromagnetic, Higgs and inflaton fields, however, we do not know whether they are composed of something more fundamental. Perhaps experiments such as those at the Large Hadron Collider will reveal that the Higgs field is not fundamental, but, for now, our understanding of Nature is too crude to be able to resolve this either way. As far as these fields are concerned, we are in a similar position to those who studied the world before they knew that atoms existed. Now, of course, we no longer regard fire, earth and water as elemental – but who knows whether or not we are really much closer than our predecessors to finding the 'true nature of things'?

Today, physicists have identified a set of what appear to be fundamental fields. The Higgs and electromagnetic fields are two, but there are also six different quark fields, six lepton fields (including an electron field and three neutrino fields), a gluon field and the weak interaction fields. All these can be described by a single mathematical framework called the Standard Model of particle physics, constructed during the course of the 1960s by Sheldon Glashow, Abdus Salam and Steven Weinberg.[1] The laws of quantum mechanics, when applied to these fields, imply that for each field there is a corresponding particle. For the electromagnetic field, the particle is the photon and for the Higgs field it is the Higgs boson. All the known types of particle in Nature are associated with the quantum behaviour of a corresponding field, and the apparent wave-like nature of electromagnetic waves can be understood as the behaviour of a large number of photons, all propagating in accord with the laws of quantum physics. The Standard Model is astonishingly successful

[1] A very brief introduction to the particles of the Standard Model can be found in the Appendix at the end of the book.

BOX 13. FUNDAMENTAL FIELDS AND THE HIGGS BOSON   P. 206

because it can describe the behaviour of all the particles we see in Nature, with the notable exception of dark matter, and the ways that they interact with each other. Armed with the Standard Model, we can explain how atoms work, the origins of radioactivity and the fusion processes that make stars burn. In fact, the Standard Model encourages us to think of the entire Universe as being built up out of a huge number of particles that hop around and interact with each other in precisely calculable ways. Only the gravitational interaction between particles is not entirely understood in this way. We understand this interaction quite well – we've spent a large part of this book talking about General Relativity – but this understanding fails when we try and work out what happens at the shortest distances, or in the proximity of huge masses at high densities such as those present at the heart of a black hole. Figuring out what dark matter is, and understanding how to compute quantum gravitational effects at very short distances, are two of the big questions that physicists are currently wrestling with.

The particle physics way of tackling these questions is the ultimate in reductionism, because everything boils down to tiny particles and their interactions. But understanding what the particles are and how they interact is not the same as understanding the richness of the Universe. Presumably, the complex behaviour of life really does emerge from the rules encoded in the Standard Model of particle physics – but it would be a fool who thought that biology is in some sense diminished by the fact. Most of us can learn the rules of chess, but very few can play the game as well as a grandmaster. But enough of those musings ... let's get back to scalar fields.

In the main text (p. 203), we said that the electromagnetic field does not pervade the whole of space in the way that scalar fields might. Roughly speaking, this means that electromagnetic fields will be largest in the vicinity of light bulbs or other sources of jiggling electrically charged particles; but, far away from such sources, they will be to all intents and purposes absent. In contrast, scalar fields can be substantial across huge swathes of the Universe. Before we get

BOX 13. FUNDAMENTAL FIELDS AND THE HIGGS BOSON   P. 207

on to the hypothesized-but-not-yet-detected inflaton field, the case of the Higgs field is worth considering in more detail.

Peter Higgs shared the 2013 Nobel Prize in physics with François Englert for their prediction of the existence of the Higgs boson, which was discovered the previous year at the Large Hadron Collider. In the 1960s Higgs and (independently) Englert – working with another Belgian physicist, Robert Brout – argued for the existence of an all-pervasive scalar field in order to solve a puzzle concerning how the particles in Nature get to have mass. Before they came up with the Higgs field (or, more correctly, the Brout-Englert-Higgs field), the mathematics governing how elementary particles behave only made sense if the particles had zero mass and zipped around at the speed of light, which is obviously nonsense. Rather than ditch the whole mathematical framework, Brout, Englert and Higgs rescued things by supposing that there is an all-pervasive scalar field, spread uniformly across the Universe. This field would interact in different ways with the various types of particle and, in so doing, would give us the impression that they have different, non-zero masses. This smart idea was central to the construction, a few years later, of the Standard Model. We aren't immediately aware of this all-pervasive scalar field because we have been living with it all of our lives: it is the backdrop to our material existence.

If this all sounds a bit like a hotchpotch of ideas, it isn't. Scalar fields can very easily fill up all of space in a process similar to how water condenses out of the air to produce dew. The equations reveal that empty space may prefer not to be empty, but rather to fill up with 'Higgs field condensation', because condensation is lower in energy than 'empty'. This might sound weird, because it seems like something is coming from nothing – but the quantum nature of the world means that empty space is never a simple thing; instead it is a seething broth of particles forever hopping in and out of existence.

In contrast, the electromagnetic field cannot condense, because it does not interact with itself. To illustrate the point, notice that the light carrying the information from a page of this book to your eyes is evidently not knocked off course

BOX 13. FUNDAMENTAL FIELDS AND THE HIGGS BOSON   P. 208

by light travelling sideways – or, put another way, photons do not interact much with other photons. In contrast, Higgs particles do interact with other Higgs particles, and this self-interaction creates the conditions that allow the Higgs field to condense out of nothing. Scalar fields do not have a monopoly on being able to condense into the vacuum; the quark and gluon fields can also condense. Today, understanding 'nothing' is one of the most interesting problems in the whole of physics.

Unlike all of the other fields we have just been discussing, the inflaton field is not well established by experiment. As we will see (p. 221), the evidence for inflation is strong – but it is not yet overwhelming. Moreover, we might entertain the idea that inflation did happen, but by means other than via the existence of a scalar field. However, again as we will see, the existence of an inflaton field can also explain the origins of the primordial lumpiness in the Universe. This is the real reason why the inflaton field is taken so seriously by so many cosmologists.

In the main text we said that the energy stored up in a scalar field can boost the expansion of the Universe. You might well ask whether the energy stored up in the Higgs field (and the quark and gluon condensates) is presently doing just that. You might also ask whether this could conceivably be the source of the cosmological constant, which we encountered in the last chapter and which is currently causing the expansion to accelerate. Not to put too fine a point on it, the situation is a theoretical disaster zone. Taken at face value, the energy stored up in the Higgs field should be causing the Universe to explode apart. That is evidently not what is happening, which means we do not understand how to account for the vacuum energy that pertains today. Our lack of understanding about why the observed cosmological constant is so small is perhaps the greatest unsolved problem in physics.

expansion, we might well ask whether the present-day cosmological constant is due to a scalar field, dubbed the 'quintessence field'. We shan't follow that idea in this book, but it is popular among cosmologists. Nor shall we worry about whether there might be some relationship between these two postulated and one observed scalar fields; the Higgs field, the inflaton field and the quintessence field. The research on these things is too speculative at the moment. Here, we want to focus on the inflaton field.

To summarize the basic idea: if the inflaton field was once present in the Universe, space could have expanded very rapidly, because the energy stored in the field acted like a large cosmological constant. This would have caused a small portion of the pre-existing space to grow rapidly to cosmic proportions, driving any particles present in the Universe before inflation to very large separations and diluting them. As time passed, the energy driving the inflationary expansion diminished until it became too small to generate any further inflation. In that way, inflation came to an end.

As this time of rapid expansion drew to a close, the inflaton field morphed from being like a still ocean, whose energy drove the accelerated expansion, into a cold gas of inflaton particles. These inflaton particles then decayed to produce a whole bunch of lighter particles, including those that populate the Universe today. The decay of heavy particles into lighter ones with the release of energy is the norm in Nature: all known particles tend to decay into lighter ones if they possibly can. We've already encountered one example in Chapter 6: isolated neutrons decay into protons in a few minutes. Top quarks decay into lighter quarks in a minuscule 10^{-25} seconds, and the Higgs particle decays in around 10^{-22} seconds. Given the fleeting nature of their existence, it is impressive that particle physicists have managed to discover the top quark and the Higgs boson, let alone measure their properties.

After the inflaton particles decayed, what remained was a Universe filled with the stuff of the Big Bang. This is very neat: the inflaton field drives a tiny patch of space into a rapid burst of expansion; then, as inflation ends, the field decays to fill the now much larger Universe with the stuff that was destined to form everything in the Universe today.

Each idea in this sequence follows from the last one in a pretty compelling way. Nothing is contrived: every piece is built on known or at least plausible physics, given what we understand about particles, fields and General Relativity. As well as doing away with the horizon and flatness problems, inflation provides a vivid description of how the Big Bang came about, because it explains where all of the particles in the Universe came from. If this were all there is to inflation, however, it would still be confined to the class of fancy scientific ideas that are doomed to remain the idle musings of theoretical physicists. We need concrete predictions that can be tested by measuring things. And this is where the idea of inflation delivers. It predicts the correct features of the ripples in the CMB and the distribution of galaxies across the sky.

As an idea, inflation took off in the early 1980s following a paper by MIT cosmologist Alan Guth, who drew attention to the virtues of working on the assumption that there was a rapid period of expansion early on in the Universe's history. Although Guth introduced a scalar field and obtained an inflating Universe, his original version of events did not lead to inflation ending with inflaton-particle decay and a resultant Big Bang. Guth realized the problem and concluded: 'I am publishing this paper in the hope that it will highlight the existence of these problems and encourage others to find some way to avoid the undesirable features of the inflationary scenario.' With this remark, he had laid down the gauntlet. Around the same time, and quite independently, Moscow-based physicists Alexei Starobinsky at the Landau Institute for Theoretical Physics and Andrei Linde at the Lebedev Institute also began to explore the idea of an inflationary phase in the early Universe. In 1982, Gary Gibbons and Stephen Hawking convened a now-famous meeting of Soviet and Western physicists in Cambridge. By the end of that year, cosmologists had not only established a working model of inflation,[4] they had also realized that quantum fluctuations in the inflaton field could generate the initial lumpiness that seeded the growth of structure in the Universe.[5]

According to the laws of quantum physics, the scalar field cannot be perfectly smooth over all of space – it cannot be a perfectly still ocean. Heisenberg's Uncertainty Principle, which is a consequence of the basic laws of quantum physics, says that the field has to fluctuate up and down slightly by varying amounts from place to place.[6] One of the implications is that 'empty' space comprises a seething broth of particles and anti-particles that pop out of nothing in pairs before coming back together and disappearing a fleet-

[4] The pioneering work was done by Andrei Linde and, independently, Andreas Albrecht and Paul Steinhardt, then of the University of Pennsylvania in the USA.

[5] In fact, the first calculations of structure formation were made before the Cambridge meeting, in 1981, by Viatcheslav Mukhanov and Gennady Chibisov, also of the Lebedev Institute.

[6] We show how the Uncertainty Principle comes about, starting from the laws of quantum physics, in our book *The Quantum Universe*.

ing instant later. In ordinary, non-inflating space these quantum effects, which are necessarily present in all fields, lead to fluctuations that average out to zero. We can glimpse them, however: the fleeting production of electron-positron pairs leads to fluctuations in the photon field that can be detected by making very precise measurements of the spectral lines emitted by hydrogen atoms. (In 1965, Richard Feynman, Julian Schwinger and Sin-Itiro Tomonaga received the Nobel Prize in Physics for working out how to perform this calculation.)

While the effect of these 'vacuum fluctuations' averages to zero in non-inflating space, something very interesting happens if the space is expanding fast enough. The Uncertainty Principle says that the particle and anti-particle pairs can exist only for a limited amount of time before they slip back into nothing. If space is inflating, however, it is possible for the particle and anti-particle to get swept so far apart that they cannot come back together to annihilate. They are carried out of each other's horizons before they can return their energy to the vacuum. If this happens, they have no option but to become real material particles. It is as if the expanding space has provided the means by which particles emitted out of nothing can escape the death-grip of Heisenberg's Uncertainty Principle. Empty space in a rapidly expanding universe glows with particles.[7] As the particles become real they add little waves to the formerly still ocean of the scalar field. In Box 14 we go into a little more detail on how it is that real particles can be produced from nothing in an inflating Universe.

---

[7] This is similar to the production of Hawking radiation from a black hole.

BOX 14. DE SITTER SPACE   P. 213

If the Universe is expanding under the influence of a
cosmological constant (and no other appreciable source
of energy), we say that the corresponding space – one of
Robertson's possible universes – is 'de Sitter space', named
after the Dutch mathematical physicist Willem de Sitter.
De Sitter space has a horizon, in the same sense that a
point on the Earth's surface has a horizon beyond which we
cannot see. On Earth, the horizon is a feature of the curved
surface. In de Sitter space, the horizon is a feature of the
Universe's accelerating expansion. To see this, imagine you
are standing in de Sitter space and watching a light bulb
recede as space expands. At some point, the light bulb will
be moving away so fast that the light it emits will never reach
you and, at that point, we say it has disappeared beyond
your horizon. Real particles and anti-particles are produced
if, after the pair has popped out of nothing, they are
subsequently swept outside of each other's horizon before
they get the chance to recombine again.

Now let's think about how this new idea impacts on our picture of the expanding Universe during the time of inflation. Consider the patch of Universe that is destined to grow into our visible Universe. If the inflaton field varies slightly from place to place across the patch then the amount of energy available to inflate the Universe will vary too, because the energy driving the expansion is controlled by the size of the field. This means that the quantum-fluctuations-made-real will cause some regions of the patch to inflate for slightly longer than other regions – and this has dramatic consequences.

We have seen that when inflation ends, space fills up with particles. Because the particles are filling up a space that has been stretched by different amounts, the density of particles will also vary from place to place. The particles will be more densely populated in regions that have not inflated as much. Crucially, this means that there will be a variation in the density of particles that is the same for all of the different types of particle. If a region has been stretched 1% more by volume than a neighbouring region, the density of photons, protons, neutrinos and dark matter particles will all be 1% lower. Cosmologists refer to this type of deviation from a smooth, perfectly uniform distribution of particles as a 'curvature perturbation' or 'an adiabatic perturbation'. Figure 8.2 illustrates a curvature perturbation in two dimensions, and the grid lines allow us to see how different regions of space have been stretched by different amounts. The corresponding variations in the density of particles created at the end of inflation are the seeds that generate the clumping of matter we see across the Universe today.

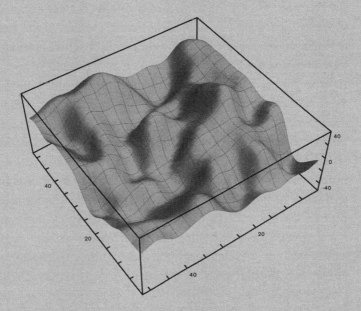

Figure 8.2 Picturing a curvature perturbation. The rippled
surface corresponds to a two-dimensional space that has
been stretched by different amounts in different places.
The space of our Universe at the end of inflation is like
a three-dimensional version of this.

To test whether this way of generating small non-uniformities in the distribution of matter and energy at the time of the Big Bang is correct, we need to understand the details of the ripples in the inflaton field, and how these lead to measurable properties in the CMB and the distribution of the galaxies. This is something we can certainly do. The way the inflaton field changes as time advances, in response to the quantum fluctuations, is best illustrated in a movie, and in Figure 8.3 we have shown stills from such a movie. The tiny ripple in the top left image represents the inflaton field when our visible Universe was about the same size as the de Sitter horizon (see Box 14, p. 213), at which time the field would have been fairly smooth over the patch.[8] For typical inflationary scenarios, the de Sitter horizon might be something like 10^{-26} metres across, which is mind-bogglingly small (it is 100 billion times smaller than a proton). The second image shows the Universe a little later on, when the original patch has been stretched and some new ripples have been created. In the first image on the second row, the ripples have stretched some more, and more new ripples have appeared. The creation of new ripples as the old get stretched continues over and over again, at a steady rate, until the end of inflation.

These stills provide a way to visualize the ripples in the inflaton field and the way they grow with time. To simplify things, they show ripples on a flat surface, like ripples on the surface of a pond, whereas in the real world they are ripples in a three-dimensional space, which are harder to visualize. We've also taken a little artistic licence to make things easier to see, by representing the region surrounding the growing

---

[8] Because the quantum-induced ripples have a size bigger than the de Sitter horizon.

patch as being completely flat. This isn't correct: it's not really flat, because it too will have ripples that were generated at times before we started the movie. We've drawn things this way because we want to track how the ripples in one part of space evolve with time, and it's easier to identify the patch we are interested in if we leave the surrounding space out of it. In reality, the first still in the sequence ought to look more like the last one, in which case our entire visible Universe might be one of the small ripples sitting on top of a stack of previously formed and stretched ones.

A particularly noteworthy feature of the pattern of ripples formed in this way is that it is what physicists refer to as 'scale invariant'. This means that we can zoom in or out of the image and it will look the same. Obviously this isn't the case the way we've drawn it, particularly in the earlier pictures – but remember, the flat regions are only there for clarity. The scale invariance becomes more evident in the final few images in the sequence. The net result of this steady production and stretching of ripples is a picture of ripples on top of ripples that looks like a bristling hive of activity on an unchanging theme. The pattern might look random, but because it is scale invariant it is also very distinctive.

The theory of inflation therefore delivers a prediction for what the Universe ought to look like when inflation draws to a close. Plate 26 shows a different way of visualizing the inflaton field in Figure 8.3, which focuses our attention on the very important prediction of scale invariance. The blue regions are where the inflaton field is smallest, and the red regions are where it is biggest. Scale invariance means that if we look at a particular patch of the image and zoom in or out, we will see patterns that are indistinguishable from each

Figure 8.3 The evolution of the inflaton field in our patch of the Universe. The top left is at a time when the entire visible Universe is 10^{-26} metres across. Time increases from top left to bottom right.

other. Since the inflaton field shapes the space that will be filled by the particles produced in the Big Bang, we might expect this scale invariance to impact on the observed temperature fluctuations in the CMB, and here a glance at the Planck photograph of the CMB is at least suggestive: Plates 25 and 26 do have certain similarities. Of course, we need to do far better than making such vague observations. We need to understand how the initially scale-invariant plasma changed as it evolved; if we can do so, we will then be able to predict what the CMB should look like in detail.

First, we need to mention that there is a subtle prediction from inflation regarding the scale invariance of the ripple pattern. The expansion rate of space must have slowed during the course of inflation, not least because inflation eventually had to end. As the expansion rate slowed, new ripples were not created quite so rapidly, and this slowing-down leads to a slight deficit of smaller-sized ripples compared to expectations based on the idea of perfect scale invariance. This point is important, because it is a pretty generic prediction of inflation. Those small deviations from scale invariance leave a measurable imprint on the CMB.

Our goal now is to work out what the observable consequences are for the CMB and the galactic structure we see in the Universe today. The theory of inflation will stand or fall based on this comparison. Perhaps surprisingly, this is not as difficult as it sounds. The whole business is rather like trying to figure out what happens if you kick a bucket of water. The kick will cause the water to be perturbed in some way and, if we know precisely how it is perturbed, we can use the equations of fluid dynamics to evolve the perturbations forward and predict how the water in the bucket will look at a later time. Inflation, like the kick, lays down the initial perturbations; the rest is dictated by equations similar to those of fluid dynamics. Instead of a bucket of water, the early Universe was a hot soup of elementary particles, but the idea is very similar. In fact, strange as it might seem, the equations for the Universe are far simpler to solve than the equations for a bucket, because of the feebleness of gravity and the small size of the initial ripples. It's not too far off the mark to say that tracking the evolution of the Universe, starting from a time when everything was compressed into a space far smaller than the nucleus of an atom, is an easier job than tracking the evolution of waves in a kicked bucket.

Figure 8.4 is possibly the most astonishing graph in the whole of physics. The data points are derived directly from the Planck measurement of the CMB. They are a mathematical representation of what our Universe actually looked like, 380,000 years after the Big Bang. The solid curve is a theoretical prediction, and you must be impressed at its accuracy; it sits bang on top of the data. You should be even more impressed when we tell you that the theoretical curve is produced from a set of initial perturbations in an otherwise smooth plasma that have their origin in curvature perturbations that are nearly but not quite scale invariant, precisely as predicted by inflation. These perturbations were evolved forward to the time of recombination, when the CMB was released. Moreover, the theoretical curve is for a universe that today contains 68% dark energy, 27% dark matter and 5% ordinary matter. This means we also have independent confirmation of all the results we established by a variety of other methods in the previous chapter! F. Scott Fitzgerald

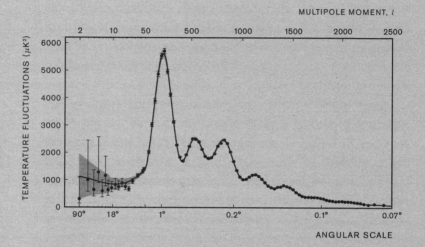

Figure 8.4 The temperature fluctuations in the
CMB as measured by the Planck satellite.

said that using an exclamation mark is like laughing at your own joke, but it is surely appropriate here. Not only does this graph support the idea that the density fluctuations in the CMB had their origin in almost scale-invariant ripples in the inflaton field, but it also provides support for our previous estimates of the amount of dark matter, dark energy and ordinary matter in the Universe, which had nothing to do with the observations of the ripples in the CMB. We'll now describe how all these key numbers describing our Universe were extracted from this single plot.[9]

First of all, we need to understand what the wiggles, or – rather more scientifically – the peaks and troughs, in the graph represent. Cosmologists refer to these as 'acoustic peaks', because they were produced by sound waves in the primordial plasma. The Universe rang like a bell, with the initial strike delivered by the curvature perturbation at the end of inflation: Figure 8.4 is a visual representation of that sound.

An analogy might be helpful. Imagine tossing a handful of pebbles onto the surface of a pond so they all land at the same time. This creates a set of disturbances in the water, which gradually evolve into a series of overlapping circles. Like water, the primordial plasma was a medium that supported the propagation of waves, and the disturbances generated by the curvature perturbation evolved in much the same way. On the two-dimensional surface of the water, the pebbled-induced waves make circles, but in the three-dimensional plasma the waves are spherical shells.

[9] To be precise, the CMB cannot be used to extract unique values for all of the key cosmological parameters. In other words, there are other combinations of parameters that also agree with the data in Figure 8.4. The remarkable thing is that agreement with the data is found using the same parameters we found in the last chapter.

Each and every point in Plate 26 acted as an initial source for an outwardly propagating spherical shell: you can think of the shells as sound waves forging through the plasma. Sound waves in air are moving variations in density, and so were the waves in the primordial plasma, but in this case it is the density of photons, electrons and nuclei that varies as the wave travels. Under- and over-dense spots in the original plasma produced density waves, while spots of average density produced no waves at all.

Crucially, because the perturbations in the initial plasma were seeded *at the same time* at the end of inflation (i.e. at the Big Bang), the spherical waves all had the *same* radius 380,000 years later, at the time of recombination. Contrast this with what would happen if you threw pebbles into a pond one at a time: the pebbles tossed in first would make waves with a larger radius than pebbles tossed in later, and the net result would be a whole bunch of circular waves, all of different radii. The peaks in Figure 8.4 are visible *because* the waves generated after inflation ended were all released at the same time.

The details of the peaks are sensitive to how the plasma was originally disturbed. We have pictured the initial disturbance – the kick of the bucket – by imagining a plasma that is created, squashed and stretched from point to point, as a result of the curvature perturbation generated by quantum fluctuations in the inflaton field. This way of initiating the density waves within the plasma is akin to plucking the string on a musical instrument by pulling it to one side and then releasing it. The waves in the plasma were initially released 'from rest'. The other way of plucking a string is to hit it, which gives it a kick and sets it moving away from an initially undisturbed position. This kind of thing could conceivably have happened to the plasma. In that case, there would not have been any initial variation in the density of the plasma. Rather, the perturbations would have been initiated by plasma flowing into or out of each region of space. This way of creating a disturbance in the plasma is known as an isocurvature perturbation. Curvature and isocurvature perturbations produce different sounds, just as plucking or striking a stringed instrument produces different sounds. Generally speaking, any kind of disturbance can be characterized as some mixture of curvature and isocurvature perturbations, and different theories for the origin of the waves in the plasma predict different mixtures. The model of inflation we have been describing is particular in selecting only curvature perturbations; it predicts that the Universe was plucked.[10] It is possible to create more sophisticated models of inflation that generate initial conditions that are a mixture of curvature and isocurvature perturbations. However, the location of the peaks in the Planck data indicates that curvature perturbations were dominant.

Hopefully, you are starting to develop a feeling for what happened in the plasma. It really is remarkably simple physics, and just goes to show that, although it may have happened a very long time ago, when the Universe was a very different place, it is not beyond our ken. In many ways, it is the world around us today that is complicated and hard to

---

[10] The 'stones thrown in water' analogy has its limitations, because really we should think of the perturbations generated by inflation more as a pressing down or lifting up the surface of the water and then releasing it from rest. The stones analogy is helpful, however, in distinguishing between the two possibilities that the perturbations were generated 'all at once and everywhere' or that they are generated progressively over time. The former case corresponds to the stones all entering the water at the same time and leads to waves of equal radii.

understand, not the Universe at its birth. The challenge facing us is to dig as much information as possible out of the primordial sound waves captured in Figure 8.4.

To do this we'll need to know more about how Figure 8.4 was actually produced. Sticking with our pond analogy for a moment, if you threw a large number of pebbles into a pond, and then took a photograph of the pond at some later time, it could be pretty hard to spot that the resulting pattern was actually produced by a series of superimposed circular waves of equal radii. This is why we cannot see anything like circles in the Planck photograph by eye. Things are made even more complicated by the fact that Planck observes a two-dimensional spherical slice through a three-dimensional snapshot of the myriad spherical waves in the plasma at the time of recombination. Fortunately, astronomers have developed techniques to sort this mess out. The result is Figure 8.4.

The first step in appreciating the details behind Figure 8.4 is to know that it is derived from something called the two-point correlation function. This function tells us how correlated the hot and cold regions are in the sky. For example, if the hot regions alone were all spaced by 1 degree, the correlation function would be large and positive at this angle. Or, if the hot regions and cold regions were all spaced by 1 degree then the correlation function would be large and negative at this angle. If there is no correlation between the temperatures at different points on the sky, the correlation function would be equal to zero. You can see that the correlation function might provide a good way to spot whether the plasma was perturbed by a whole bunch of superimposed spherical sound waves, and that it could inform us of their radii. Indeed, this is the case, as the positions of the peaks in Figure 8.4 are related to the radius of the spheres.[11] In Box 15 (pp. 233–7) we give a much more detailed description of the physics responsible for Figure 8.4. It is particularly important to emphasize that if the spherical shells were not all of the same size, there would not be any peaks. The mere existence of the peaks in Figure 8.4 informs us that the structure in the Universe was laid down once and for all at the time when the sound waves were first launched. Inflation provides a means to orchestrate this grand opening to the Universe.

There is even more information hiding in the details of Figure 8.4. The photons received by Planck all began their journey from the surface of last scattering almost 14 billion years ago, as illustrated in Figure 8.1. This means that we are looking from a vast distance at a pattern built up from lots of spherical waves. The size of the spheres we see in the CMB from our vantage point on Earth therefore depends on two separate things. Firstly, the observed size obviously depends on the actual size of the spheres at the time when the CMB photons were released from the plasma, which is determined by the speed at which sound waves moved through the plasma. And, secondly, their observed size depends on the distance to the surface of last scattering, which is given by the expansion history of the Universe: that is, the more distant the surface of last scattering, the smaller the spheres will appear. As we've seen, this distance depends on how much the Universe has expanded during the time the photons have been travelling to Earth, and this is related to the amount of

[11] For the more mathematically inclined, Figure 8.4 is closely related to the Fourier transform of the correlation function.

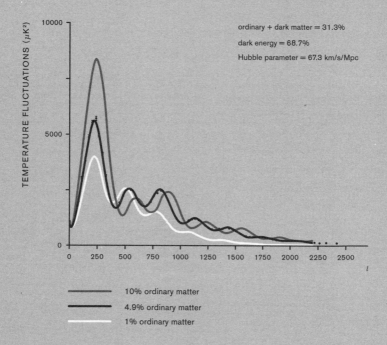

ordinary + dark matter = 31.3%

dark energy = 68.7%

Hubble parameter = 67.3 km/s/Mpc

10% ordinary matter

4.9% ordinary matter

1% ordinary matter

Figure 8.5 The temperature fluctuations in the Cosmic Microwave
Background are highly sensitive to the key numbers governing the composition
of the Universe. Notice how much the curves change as we vary the amounts
of ordinary matter, dark matter and dark energy. In the right-hand graph, the
grey and white curves have less than 100% of the critical energy density,
which means they correspond to a non-flat geometry for the Universe. In all
cases, except for the white curve in the right-hand graph, the typical size of the
initial ripples arising after inflation is the same. For this white curve, the initial
ripples were chosen to be smaller, otherwise the curve would be way too big.

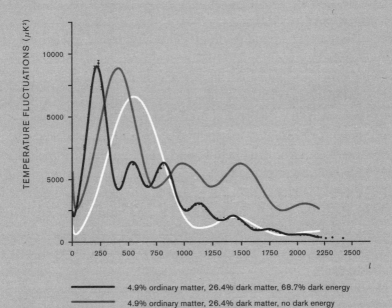

TEMPERATURE FLUCTUATIONS (μK²)

10000

5000

5000

5000

0   250   500   750   1000 1250 1500 1750 2000 2250 2500

l

—— 4.9% ordinary matter, 26.4% dark matter, 68.7% dark energy

—— 4.9% ordinary matter, 26.4% dark matter, no dark energy

—— 30% ordinary matter, no dark matter, no dark energy

matter, dark matter and dark energy in the Universe. The fact that the observed size of the spherical sound waves depends on the expansion history of the Universe is one reason why Figure 8.4 can be used to help us determine how much and what types of matter there are in the Universe. If you want to know more, then take a look at Box 15.

We can get a good feeling for just how sensitive Figure 8.4 is to the composition of the Universe by showing some curves for what it would look like if we changed the amounts of matter in the Universe. The left-hand graph in Figure 8.1 shows how the theoretical prediction changes as the relative amounts of ordinary matter and dark matter are altered (the sum of the two being held fixed). If you worked through the box then you might be able to figure out why the curves look as they do. The right-hand graph emphasizes how hard it is to make the theory agree with the data, and therefore how impressive it is that the sweet spot of near-perfect agreement occurs using the same numbers we harvested in the last chapter using other astrophysical observables. If there was something wrong with our understanding of this vast swathe of physics, it is very hard to see how we would get such beautiful agreement.

We've been focusing a lot on the way that the initial perturbations in the primordial plasma led to the microwave background, but, as we said, the same initial perturbations also led to the formation of the galactic structures we see in the SDSS map (Plate 22). The fact that the simulations like those in Plate 24 agree with observations is strong evidence that the model is good. There is also a very striking way to see the imprint of those sound waves in the primordial plasma.

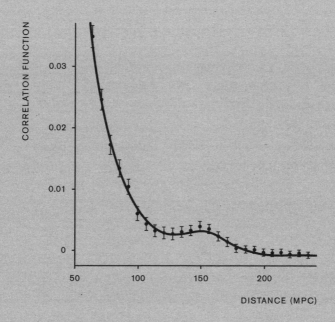

Figure 8.6 The two-point correlation function, which shows
that galaxies are mainly found to be close together (hence the
rise at low distances). Far more interesting is the little bump at
150 Mpc. This is exactly what would be expected if the galaxies
formed preferentially on over-dense regions corresponding to the
expanded shells, which are the imprint of the earlier plasma sound
waves. The solid curve is the prediction using the same parameters
as were used to produce the curve in Figure 8.4.

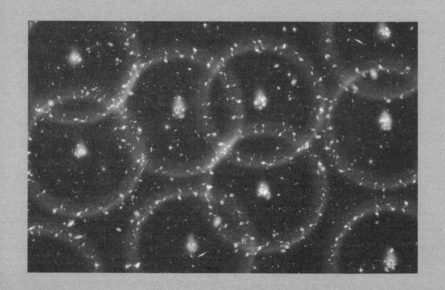

Figure 8.7 Baryon Acoustic Oscillations mean that there
is a higher than average probability of finding pairs of
galaxies that are separated by around 150 Mpc.

We have seen that, at the time of recombination, the plasma tended to be over-dense in spherical shells as the sound waves travelled outwards from initially over-dense regions. This means that, at the time of recombination, protons tended to have a higher-than-average chance of being in these spherical shells. These are the same protons that, much later in the history of the Universe, formed the hydrogen gas that collapsed to form galaxies. The original spherical shells, slightly richer in protons than the surrounding regions, grew with the expansion of the Universe, and (using the Friedmann equation again) they had a radius of around 150 Mpc when the first galaxies were formed. The situation is represented schematically in Figure 8.7. Here is a direct prediction: if we play the same game as we did for the CMB, and construct a correlation function, this time for galaxies, then we should see that there is a slight tendency for pairs of galaxies to be separated by a distance of 150 Mpc. Figure 8.6 shows the observational data and, quite remarkably, we can see that the prediction is correct; there is a peak in the correlation function at 150 Mpc. Perhaps we really do understand the evolution of the Universe, from a time before the Big Bang all the way through to the present day.

BOX 15. SOUND WAVES IN THE PRIMORDIAL PLASMA   P. 234

We want to understand how the peaks in the Planck graph relate to the radius of the spherical shells. To do this, we are going to introduce an entirely different, but totally equivalent, way of thinking about waves in the plasma (or in a tank of water, or in the air). Plate 27 shows how it is possible to think of a pattern of disturbances as being due to plane waves in combination. Plane waves are special waves, like the three to the left of the equals sign in Plate 27. It is clear why they are called plane waves: they look like stacked planes. The light and dark regions correspond to places where the wave is big and small. The waves we have in mind correspond to variations in the density of the plasma in the early Universe – but, on a more down-to-earth level, they could be variations in the density of the air in a room as a sound wave travels through it. The three plane waves in Plate 27 just happen all to have the same wavelength (the distance between successive planes, indicated by λ in the figure) and they are all arranged at right angles to each other. This is why the resultant pattern obtained by adding them together is so regular. But there is nothing to stop us building patterns by adding together plane waves of different wavelengths and orientations. It doesn't take too much imagination to appreciate that it might be possible to construct any particular set of disturbances in the plasma by adding together a large number of plane waves. The idea of building ripples in the plasma by adding together a bunch of plane waves is simply for our convenience; it is a means to an end. The virtue of thinking like this becomes evident when we consider what happens to one particular plane wave as time advances.

Plate 28 shows a slice through a plane wave, and the sinusoidal wave below it indicates how the density in the plasma varies along the wave. We are imagining that this particular wave is one of many plane waves that must be added together to describe the plasma as it is delivered to us at the end of inflation. At time zero, this wave is released into the plasma. By this we mean that, at time zero, the plasma was squashed and squeezed in the pattern of a plane wave. A short time after, the particles in the plasma will start

BOX 15. SOUND WAVES IN THE PRIMORDIAL PLASMA   P. 235

to move as they are pushed away from the higher-density regions where the pressure is greatest. This happens across the entire wave, the net result being that we can think of what happens as the formation of two waves, each half the size of the original wave. One of these waves moves to the left at the speed of sound; the other moves to the right. We must add together the two waves to ascertain the net effect. This is illustrated in the figure for two moments shortly after time zero. Notice that the resultant wave is the same shape as the original one; all that is changing is its size. In other words, the original wave simply oscillates in size, first getting smaller and then growing back again. For example, there will be a time when each half-wave has travelled exactly one quarter of a wavelength. When that happens, the peaks in one wave will coincide with the troughs in the other and the two will exactly cancel each other out. At this moment, the plasma will be perfectly uniform. However, this is a fleeting moment because the particles in the plasma are moving and they overshoot perfect uniformity. At the time when the two waves have travelled exactly half of one wavelength, the initial plane wave will be inverted; initially high-density regions will now be low-density regions, and vice versa. The net result of all of this is that, if you were located somewhere in the plasma, the plasma in your vicinity would oscillate periodically between high and low density. The time between successive peaks in intensity is the time it takes those two half-waves to travel one full wavelength (then they will have exactly passed through each other once), i.e. the period of oscillation of the standing wave is equal to $T = \lambda/v$, where v is the speed of sound in the plasma (the speed at which waves travel through it).

The Planck image is a snapshot of the plasma at the time of recombination, as viewed from the Earth. To obtain the behaviour of the plasma, we must combine the effect of many plane waves of different wavelength. Using what we have just learned, we can determine which standing waves will correspond to the biggest disturbances in the plasma at recombination. Although all the standing waves were originally as big as they can be (i.e. all waves started out by

BOX 15. SOUND WAVES IN THE PRIMORDIAL PLASMA   P. 236

diminishing in size – the Universe was 'plucked', as illustrated in Plate 28), each oscillates with a different time period, and only waves of certain wavelengths will be big again at the time of recombination. Specifically, the biggest waves will be those that have undergone either an integer or a half-integer number of oscillations at recombination. Standing waves that have undergone an integer number of oscillations look exactly like the original wave, while those that have undergone a half-integer number of oscillations look pretty much the same, except that the high-density regions and low-density regions are interchanged. Waves that have undergone a quarter-integer number of oscillations will not produce any disturbance at the time of recombination. This means that the biggest waves at recombination will have a wavelength equal to $2vT/n$, where T is the time of recombination and n is an integer ($n = 1, 2, 3, \ldots$).[1]

Now, we can make the link to Figure 8.4, the graph mapping out the ripples in plasma at the time of recombination. The mathematical processing of the raw Planck photograph that leads to this figure is specifically designed to help us see which plane waves are big at the time of recombination. To be precise, the angle marked on the horizontal axis is the angle between successive peaks in plane waves that were oriented across the line of sight at the time of recombination. This means that waves of shorter wavelength contribute to Figure 8.4 at smaller angles. The first peak occurs at an angle of close to 1 degree, which tells us that $1/360 = \lambda/2\pi R$, where R is the distance the CMB photons have had to travel to reach Earth. Given what we just did in the last paragraph, we know that $\lambda = 2vT$ (corresponding to $n = 1$), which means that the position of the first peak tells us that the speed of sound through the

---

[1] Because the Universe is expanding all the time, we need to be a little careful in defining times and distances. If we measure distances (like the wavelength λ) using rulers that stretch with the expansion of the Universe (the grid lines on Figure 8.2 could be used to measure distances in this way), then the time T should be what cosmologists call the conformal time, which is the distance an unhindered beam of light would travel in that interval of time divided by the speed of light.

BOX 15. SOUND WAVES IN THE PRIMORDIAL PLASMA   P. 237

plasma is equal to $\pi/360 \times (R/T)$. The values of R and T are influenced by the amount and type of the stuff that is causing the Universe to expand, because that is what controls when recombination happens (which fixes T), and how far the CMB photons have to travel before they reach our telescopes (which fixes R). You should now be able to appreciate why the CMB graph in Figure 8.4 can be used to extract information on how fast the plasma waves travel (which is governed by the composition of the medium in which they exist) and how much the Universe has expanded since the CMB photons started their journey (because increasing the amount of expansion increases the distance the photons must travel to reach Earth, which in turn reduces the angle any particular plane wave subtends on the sky).

So far, our focus has been on the positions of the peaks, but the heights of the peaks contain information too. The general trend is for the peaks to diminish in height as the angle decreases, which means that the standing waves of smaller wavelength were less prominent than the longer-wavelength ones at recombination. This is because the photons bounce around inside the plasma and this produces a blurring of the standing waves, that is, peaks and troughs are smeared out by a distance determined by how far a typical photon travels between successive collisions with the electrons and protons. If this distance is larger than the wavelength of a standing wave, then that wave will be washed out. The decline in size of the acoustic peaks is because of this gradual dissolution of the standing waves. This logic also implies that there will be less damping of the short-wavelength waves if there are more electrons and protons in the plasma, because this is what controls the typical distance a photon travels between collisions: if there are very few charged particles, the original waves will dissolve as the photons simply stream away from the regions of high density. That's nice, then: the heights of the peaks are sensitive to the charged particle density in the plasma.

The cherry on the cake is the fact that careful inspection of the graph reveals that the odd-numbered peaks are

BOX 15. SOUND WAVES IN THE PRIMORDIAL PLASMA   P. 238

boosted relative to their adjacent even-numbered peaks. Given what we said in the last paragraph, we might expect the heights of the peaks to fall steadily as the angle decreases, corresponding to shorter wavelengths. However, this isn't quite what happens. This failure of the peak heights to steadily reduce is particularly visible if we look at the second and third peaks, which are almost the same size. This boosting of the odd-numbered peaks is a direct consequence of the fact that the plasma is oscillating in a background of dark matter, which tends to gravitationally attract the charged particles towards it. This has no effect on the standing wave pattern after each full oscillation, because each standing wave always rebounds back to the same starting point. But it does have an effect on how deeply the plasma is able to compress before it rebounds back. Increasing the density of the charged particles in the plasma causes the half-integer oscillations to be more intense, which boosts those odd-numbered peaks.

We have only touched on the most important ideas here – but it is deeply satisfying that every feature of the CMB graph can be accounted for with basic physics.

9.
OUR
PLACE

We have travelled a long way from those idle contemplations on the beach at Ogmore-by-Sea. We have followed in the footsteps of many ordinary people, who took their musings seriously enough to act on them and spent their lives trying to understand. Standing on the seashore, Mike Seymour noticed something that piqued his curiosity. That inquisitiveness, that spirit of enquiry, is the thing that leads us to become scientists and take seriously the questions we have about the world. Back in the eighteenth century, Henry Cavendish's urge to quantify gravity and measure the mass of the Earth unlocked the power of Newton's laws. He dared to imagine, but he also dared to conduct an experiment that required painstaking effort, meticulous craftsmanship and integrity. He undertook to question every aspect of his measurements and was never going to allow himself to be swayed by his own preconceptions or by pressure from his rivals. Today, big advances are often made by large collaborations of people, but the motivation is just the same. The Planck satellite project and the Large Hadron Collider project at CERN involve thousands of scientists rather than one, but the only real difference is that, because of the complexity of the measurements they need to make, a bunch of enthusiasts had to club together to build the experiments and analyse the data. The powerful desire to understand brings people together from all over the world in these large international collaborations. National boundaries dissolve and are replaced by an unfettered spirit of co-operation. This spirit is one of the great joys of science, and because of it great things happen.

Throughout this book we have shown how people have gone about measuring things and, in so doing, demonstrated a way to accumulate reliable knowledge. Our thoughts have led us outwards from the Earth to the stars, and to the galaxies beyond. Eventually, we have reached a point where we can seriously consider the origins of the Universe. If we lift our gaze from the parochial and anthropocentric, the cosmos awaits us in humbling, awe-inspiring glory.

It is staggering to suggest that the entire observable Universe came from a subatomic-sized patch of space, but the theory of inflation accounts for how such a patch evolved into the Universe we inhabit. The theory leads to a Big Bang and generates a pattern of initial perturbations that has been confirmed by precise measurements of the Cosmic Microwave Background and the clustering of galaxies. What's more, inflation predicts the existence of primordial ripples in spacetime that may be observable today as gravitational waves: the measurement of these would, for many people, be the clinching piece of evidence confirming that inflation really did happen. Work towards that goal is underway.

Inflation certainly provides a vivid account of the origins of the observable Universe. It is a daring theory with high ambition. However, what we have presented so far is positively prosaic when compared to inflationary theory's most astonishing prediction: that our visible Universe might be just one of a possibly infinite number of universes.

In the last chapter, we saw some movie stills of our patch

of Universe as it evolved during inflation. Our particular patch started out tiny, and there is no reason to think that it was anything other than one small part of a much bigger space, which may be infinite in extent. We chose to start the movie sequence when the observable Universe was big enough to start developing ripples but, presumably, the rest of the space had already been inflating for some time by then, and we joined the story part way through. Over the following few million million million million million millionths of a second, the inflaton field caused our patch to grow to the size of a melon, before the exponential expansion came to an end as the energy driving it ran out.

Let's now consider what was happening in the rest of the inflationary Universe. We have said that the inflaton field gradually faded as it drove the expansion of our patch and ultimately decayed away completely at the Big Bang. This implies that the inflaton field would have been bigger in the past, and the acceleration of the expansion of space more rapid. This is an 'on the average' statement, because the size of the inflaton field actually varied from place to place because it had quantum-induced ripples in it, and the places where the field was bigger than average would have inflated more rapidly. In our patch, the ripples were small; their main effect was to sow the seeds that gave rise to the perturbations in the CMB and the network of galaxies that fill the sky today.

Now, here is a crucial piece of information: when the inflaton field was larger, the ripples were larger too. As we go back in time, we see the inflaton field fluctuating more and more wildly; the tiny ripples in a still ocean were preceded by huge waves on a stormy sea. We can think of the situation another way, using an idea from the last chapter. During inflation, space glows with particles, and the glow becomes more intense as we look further into the past and the acceleration of the space increases.

We have said that the size of the inflaton field steadily falls as inflation progresses and that when it falls below a certain level inflation ends. At early enough times, however, this was not necessarily the case, because the ripples in the inflaton field were so dramatic that some regions received fluctuations large enough to cause the inflaton field to *increase* in size, notwithstanding the dilution during the expansion. In those regions, inflation would have sped up instead of slowing down. If regions like this are created at a sufficient rate, there will always be some parts of space that are inflating and inflation carries on for ever. There will still be regions where the field is not so big, and inflation will end in these regions, giving rise to Big Bangs. Our observable Universe emerged from one such region. In this scenario, the entirety of space contains bubbles of universe where inflation has ceased, isolated from each other by regions where inflation is ongoing. This is known as the inflationary Multiverse.

In the Multiverse, universes are being created out of nothing, and this seems wrong. For one thing it appears to violate the law of energy conservation, which states that the total amount of energy in a closed system does not change. This is not a problem in General Relativity; there is nothing wrong with the idea of an expanding space in which the total energy carried by its contents changes. This idea is manifest in our observable Universe; as space expands, the photons it contains are redshifted. This is the effect that we used to measure the rate at which space is expanding in Chapter 6. Quantum theory tells us that the energy of a

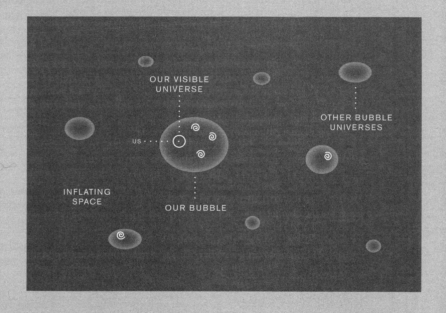

Figure 9.1 The Multiverse. Note that the scale here is wrong: the typical distance between bubbles is vastly bigger than their sizes. Our type of universe might be relatively rare in being old enough and big enough to support the evolution of intelligent life.

photon is inversely proportional to its wavelength, so, as the photon's wavelength is stretched by the expansion of space, its energy falls. This is why the Cosmic Microwave Background temperature is only 3 kelvin today, even though it was emitted at a temperature of 3000 kelvin at the time of recombination. Evidently, the total energy of an expanding universe containing only photons falls. The opposite is true for a universe expanding due to a cosmological constant. In this case, the energy density in space stays constant even though space is expanding, which means the total energy in the universe increases.

A brief recap is in order. We introduced the inflaton field in order to solve the horizon and flatness problems in a manner consistent with the laws of particle physics as we understand them today. All we initially wanted to do was to create a theory capable of describing a rapid expansion in the early Universe, and just look where it has taken us. Inflation's logic has pushed us into an explanation for the origin of the Big Bang and the distribution of galaxies in our Universe, not to mention the detailed structure of the CMB. Now we discover that the same theory appears to predict that this whole process probably happened over and over again, littering the Multiverse with a vast number of bubble universes, of which ours is just one. It may be that we will never be able to check experimentally if there really is a Multiverse, although we may be lucky. Some cosmologists have speculated that the bubble universes might collide with each other early in their evolution, and that this may have left a faint signature waiting for us in the CMB. Perhaps there are other, as yet unimagined, theoretical developments or measurements that might allow us to infer the existence of the Multiverse.

There is another popular idea in theoretical physics that, at first sight, has little to offer us in our contemplations of the Multiverse but which, on closer inspection, turns out to add a remarkable metaphysical twist. For a large part of our scientific lives, String Theory has been dominant in the holy-grail quest for a Theory of Everything, bringing together quantum theory and General Relativity into a beautiful and consistent whole. The idea is that everything in the Universe is constructed out of tiny loops and strands of vibrating 'string'.[1] The typical size of a piece of string is around 10^{-35} m, which is a hundred billion billion times smaller than a proton – this smallness explains why we might have hitherto been fooled into thinking that everything is made out of particles. String Theory gathered momentum in the mid-1980s, especially following the work of Mike Green, then of Queen Mary College in London, and Caltech's John Schwarz. This was when String Theory started to be taken seriously as a possible Theory of Everything – theorists realized that not only did String Theory contain General Relativity, it also gave rise to physics resembling the well-established Standard Model of particle physics. String theorists began to dream that there may be only one logically possible theory, and that this theory would predict all of the laws of physics as we experience them. It would explain the values of all the particle masses, the cosmological constant and the strengths of the forces. The seductive idea is that the Universe we live in is the only logically possible universe, and that underlying everything is a perfect, unique mathematics. Sub-

---

[1] The possibility of strings is a natural extension of the idea of a particle. While particles are zero-dimensional objects, i.e. points in space, strings are one-dimensional objects, i.e. lines in space.

sequently, a good deal of progress was made towards showing how the laws of physics we observe today *could* emerge out of String Theory, but nobody managed to find the holy grail – the one theory to rule them all.

In practice, String Theory is complicated, because its mathematical consistency demands the strings should vibrate in a ten-dimensional spacetime. Since we only experience four dimensions, the other dimensions need to be hidden from us in some way. One way to do that is to have the extra dimensions curl up into tiny geometrical shapes at every point in our space; so tiny that we simply can't see them. A familiar analogy would be to imagine a hosepipe at the bottom of a long garden. From sufficiently far away, it looks like a one-dimensional line, even though it is in fact a two-dimensional surface rolled up into a cylinder. In such a picture, our experience of the world is to be seen as a coarse, 'large distance' one, and the laws of physics we encounter today are to be regarded as emergent 'low energy' approximations to the true theory. Naturally enough, scientists took great encouragement from the fact that laws of Nature resembling those we know emerge from String Theory.

In the early years of the twenty-first century, however, the optimism was tempered by the gradual realization that String Theory didn't predict a unique set of low-energy laws. Instead, it seemed to predict a grand array of possible universes, each with a different set of emergent laws that are distinguished from each other by the different ways that those extra dimensions in space are curled up. This prompted Leonard Susskind, of Stanford University, to introduce the idea of a String Landscape. You might picture a vast landscape of hills and valleys, stretching off as far as the eye can see, with every valley corresponding to a different possible universe with different low-energy physical laws. This might have been fine if there were only a handful of possibilities, but it became clear that there could be as many as 10^{500} low-energy manifestations of String Theory, and we appear to live in a universe described by one of them. In a remarkable about turn, many string theorists went from seeking a unique Theory of Everything to exploring a theory that has the potential to enumerate a near-infinite variety of possible worlds.

At first sight, this may seem like a terrific disappointment – are there really a vast number of possible universes? And if so, how can we possibly figure out why we live in the one we do? These questions echo those articulated by Gottfried von Leibniz, in his 1710 work *Théodicée*. Leibniz considered the idea of a vast number of possible worlds and believed that we inhabit 'the best of all possible worlds'.[2] However, the ambiguous plethora of the String Landscape raises a fascinating alternative – might all of these possible universes actually be realized in the vast Multiverse of eternal inflation?

The central idea of eternal inflation is that there will always be portions of space that are accelerating rapidly because of unavoidable upward fluctuations in the inflaton field. In these regions, the larger value of the inflaton field causes the energy density to increase and, at very high energy densities, we need something like String Theory to compute what happens. Crucially, at such energy densities, it seems

---

[2] More recently, the idea that 'the world we are part of is but one of a plurality of worlds' is a key idea in the 'modal realism' advocated by the late David Lewis in his 1986 book *On the Plurality of Worlds*. Max Tegmark's book *Our Mathematical Universe* (2014) provides a thought-provoking introduction to how modern physics leads to what he refers to as the four levels of the Multiverse.

to be possible for the laws of physics to be re-set, as the Universe makes a transition from one valley in the String landscape to another. Roughly speaking, the tiny, curled up extra dimensions unravel and re-adjust; it is as if space is melting and re-crystallizing in a different configuration. This mechanism for moving about the String Landscape makes it possible for different bubble universes in the Multiverse to have different low-energy laws of physics.

Viewed this way, String Theory and the inflationary Multiverse fit neatly together. String Theory allows for the possibility that there are very many different ways of arranging 10-dimensional spacetime to produce different emergent laws of physics, and the Multiverse provides a mechanism for realizing them all. According to the theory as we have just described it, although different bubble universes will generally exhibit different low-energy laws, they are all still governed by the same overarching laws of String Theory. Low-energy observers like us are trapped in one particular valley in the landscape, and that is what shapes our view of the Universe. Other vantage points in the landscape would correspond to vastly different physics: different elementary particles, different forces of Nature, and even different dimensionalities of space. Across the Multiverse, this means there will be universes with stronger gravity and vastly larger cosmological constants, universes with no atoms or stars, universes filled with black holes and universes that are almost, but not quite, the same as ours. In all likelihood, there will be vastly more universes than there are atoms in our observable Universe.

If what we just described is really the way things are, then the implications for how we view our place in the cosmos are clearly profound. We must add the caveat, though, that the evidence for inflation itself is not yet absolutely compelling, and we have no firm evidence for the validity of String Theory or the Multiverse. Nevertheless, the case for the Multiverse is not plucked out of the head of an imaginative dreamer. It is built on a chain of reasoning that is more or less compelling, depending on who you ask. And, of course, the frontiers of science must always lie in the realm of speculation – collecting evidence can be a lengthy and difficult task. One of the major challenges facing scientists today is to uncover the evidence that is needed before we can be confident in the theoretical ideas presented in this chapter.

The question of the origin of our Universe is obviously of immense cultural significance. Are we the result of an intelligent creator? Is there a reason for the existence of the Universe? At first sight, modern physics offers some of the strongest evidence in favour of the idea that the Universe was designed. The fundamental laws of Nature are astonishingly compact, powerful and beautiful. The Standard Model of particle physics, which describes how all of the particles in the Universe interact with each other, is possessed of a high degree of symmetry. When you work with the mathematics behind the physics it is impossible not to be touched by its elegance; the equations make snowflake-beautiful patterns that encode the laws of Nature. It is as though a brilliant mathematician set up the Universe. This 'unreasonable effectiveness of mathematics'[3] might be invoked to appeal to a

---

[3] This is the title of theoretical physicist Eugene Wigner's 1960 essay that ends, 'The miracle of the appropriateness of the language of mathematics for the formulation of the laws of physics is a wonderful gift which we neither understand nor deserve.' It echoes Einstein's quote: 'The most incomprehensible thing about the universe is that it is comprehensible.'

higher intelligence, as is done by the Reverend John Polking-horne when he states that[4] '[t]he world that science describes seems to me, with its order, intelligibility, potentiality, and tightly knit character, to be one that is consonant with the idea that it is the expression of the will of a Creator.' Polking-horne knows the mathematics well; he was a particle physicist and professor in mathematical physics at the University of Cambridge before joining the priesthood. Even though many theoretical physicists do not go so far as to invoke the idea of a Creator, they are still deeply affected by the remarkable beauty inherent in the fundamental equations in physics.

The argument for a Creator also appears to be bolstered by another remarkable aspect of the natural world: the laws of physics seem to be perfectly adjusted in order to produce a Universe that is hospitable to life. To apply the fundamental laws of physics, as encoded in the Standard Model and General Relativity, it is necessary to first specify the values of around thirty numbers, which include the strengths of the forces, the masses of the particles and the size of the cosmo-logical constant. Only once these have been fixed can the equations be used to predict the outcomes of experiments and observations. Changing these numbers, often by just a few percent, gives rise to theoretical universes that have no chance of supporting life. It is very easy to end up with a theory in which stars never form, or burn out in millions rather than billions of years, leaving no time for biological evolution. The strengths of the forces seem particularly well adjusted to avoiding these disaster scenarios. It is also very easy to end up with a theory in which the chemical elements

[4] In *One World: The Interaction of Science and Theology* (2007).

– the necessary building blocks of all complex structures – do not exist in anything like the form we know. The periodic table is a delicate balancing game, and it appears to be extremely difficult to pick values for the strengths of the forces and the particle masses such that the heavier elements, including carbon and oxygen, are produced in stars and remain stable against radioactive decay. The expansion rate of the Universe is also 'just so': it would be easy to make a universe in which matter never clumped together, which is what happens if the cosmological constant is too big. There would also be no stars or galaxies if there were too little dark matter or too much light. Even the way that supernovae explode, scattering the heavy elements necessary for life across interstellar space, would be significantly affected if the weak nuclear force was just a little weaker or stronger than it is. Taken at face value, our Universe has a bespoke feel to it. In the words of the great theoretical physicist Freeman Dyson: 'As we look out into the Universe and identify the many accidents of physics and astronomy that have worked together to our benefit, it almost seems as if the Universe must in some sense have known we were coming.'

Obviously the numbers that characterize our Universe – such as the particle masses and the force strengths – must describe a universe fit for life, because we exist. That is not at issue. What is at issue is the nature of the mechanism for selecting those numbers: did the mechanism possess a creator-like foresight or not? The Multiverse says 'not', because it delivers an almost inconceivably rich variety of bubble universes that manifest all of the possible laws of physics. It does this randomly, with no foresight, and it guarantees the existence of an apparently unique universe such as ours.

Viewed this way, our existence is inevitable – along with the existence of every other conceivable universe.

The Multiverse idea clearly undermines the argument for a Creator based on the fine-tuning of the laws of Nature. However, it does not quite undermine the argument for what we might call a Creator-Mathematician. String Theory potentially provides a very beautiful mathematical construction that overarches the Multiverse, and the origins of that theory remain mysterious. Perhaps some intelligence is responsible. But if you would still like to posit a Creator, the Multiverse idea paints a striking picture of their methods. It seems that the architect of our Universe set about their task by creating universe after universe. For each universe, the laws of physics were selected at random – as if by rolling dice. This kind of thing is reminiscent of what scientists do when they want to simulate systems they do not understand and want to understand better. They put the equations that generate the system onto a computer and allow the system to evolve while they watch what happens, often choosing the key numbers that determine the evolution of the system at random to generate a wide diversity of outcomes. The simulations of chunks of universe we presented in Chapter 8 were created like this.

Today, the cosmologists responsible for those simulations are hampered by insufficient computing power, which means that they can only produce a small number of simulations, each with different values for a few key parameters, like the amount of dark matter and the nature of the primordial perturbations delivered at the end of inflation. But imagine that there are super-cosmologists who know the String Theory that describes the inflationary Multiverse. Imagine that they run a simulation in their mighty computers – would the simulated creatures living within one of the simulated bubble universes be able to tell that they were in a simulation of cosmic proportions?

Science has made an astonishing amount of progress over the past 500 years. Each new generation of scientists has benefited hugely from the often herculean efforts of those who went before, and today we find ourselves in the very privileged position of being able to contemplate and to compute what happened at the birth of our Universe. The process of learning new things is not mysterious, but there is romance in the endeavour. There is something truly wonderful about the determined pursuit of seemingly insignificant details. At its heart, science is about connecting with the world; it is a living celebration of the Universe. It is about reaching out into the unknown and exploring the uncharted landscape of ideas. We are part of the greatest of mysteries, and, for us, that is enough.

APPENDIX

The following is a list of some basic maths and physics that may be helpful.

POWERS OF 10

In cosmology and particle physics we often encounter numbers that are either very big or very small. To help write such numbers we use exponential notation. For example, 1 million, which is equal to 1,000,000, can be written 10^6. This should be read as '10 to the power of 6', which means it is equal to 10 multiplied by itself 6 times. Tiny numbers are written with negative powers, so one billionth, which is equal to $1/1,000,000,000 = 0.000000001$, is written 10^{-9}.

UNITS

Very often we deal with quantities that carry units. The simplest example might be a distance, such as 1 kilometre = 0.621 miles. As in the case of kilometres and miles, we always have the freedom to choose the units we want to use to measure something. Although metres and kilometres are convenient units for stating the typical distances we encounter in our daily lives, they are not very convenient in cosmology and particle physics. More commonly we will want to use light years, megaparsecs, ångstroms and nanometres. These can all be converted into metres as follows:

1 light year = 9.46×10^{15} metres
1 megaparsec = 1 Mpc = 3.26×10^{6} light years
1 nanometre = 10 ångstroms = 10^{-9} metres
1 femtometre = 10^{-15} metres

Likewise, in everyday circumstances it makes sense to meas-ure energies in joules (e.g. a 20 watt light-bulb radiates energy at a rate of 20 joules per second), but when discussing particles of atom-size and smaller it makes more sense to use the electronvolt (denoted eV), which can be converted into joules using:

1 eV = 1.60×10^{-19} joules

We will often abbreviate units, e.g. 1 nm = 1 nanometre or 1 km = 1 kilometre or 1 mega-electronvolt = 1 MeV.

Numbers that carry with them an associated unit are called dimensionful numbers, and it is common to want to com-bine dimensionful numbers by multiplying or dividing them. The simplest example of this is when we take a dis-tance and then divide it by a time in order to get something that is a speed. For example, a car moves 100 km in 2 hours, therefore its speed is 100 kilometres divided by 2 hours = 100 km/2 h = 100 × 1 km/2/(1 h) = 50 × 1 km/1 h = 50 km/h. Obviously you could see that the car moves at 50 km per hour straight away – but we chose to show the various intermediate ways we could have written the ratio because simple manipulations like these are sometimes per-formed in the text.

We encounter two particularly elaborate dimensionful numbers in the book: the Gravitational constant, $G = 6.67$ m^3/s^2/kg and the Hubble constant, $H = 68$ km/s/Mpc. These units might look abstract, but in fact they are easy to comprehend. For example, divide G by a distance squared and then multiply it by a mass and you end up with a number with the units of acceleration (m/s^2): if the distance is the radius of the Earth and the mass is the mass of the Earth, then the corresponding acceleration is the acceleration due to gravity, for an object dropped close to the Earth's surface. Likewise, the units of the Hubble constant tell us that a galaxy 1 Mpc away recedes from the Earth at a speed of 68 km/s.

Occasionally we make use of elementary algebra. For example, the calculation of the speed of the car above could be carried out using the formula $v = d/t$ where $d = 100$ km and $t = 2$ hours, to give $v = 50$ km/h. A formula can be transformed into another equally valid one by performing the same operation on each side of the equals sign, e.g. $v \times t = vt = d/t \times t = d$. In this case, we multiplied both sides of the equation by t to deliver an equation telling us that the distance d is equal to the product of the speed, v, and the time, t. Notice how we can write a product of two numbers either using an explicit multiplication symbol ($v \times t$) or, more simply, as vt. We always denote division by a backslash symbol (e.g. $d/t \times t$ means 'd divided by t multiplied by t', which is simply equal to d).

ELEMENTARY PARTICLES
Atoms are the building blocks of the ordinary matter we encounter on Earth. They are approximately an ångstrom

across, and most of their mass resides in a tiny central nucleus built from protons and neutrons. Protons are only about 1 femtometre in diameter and they carry positive electric charge. Orbiting around the nucleus are the electrons, which carry negative electric charge such that the entire atom is neutral. The simplest atom is called hydrogen and it consists of a single proton and a single electron – it is the most abundant type of atom in the Universe. The way that the electrons are arranged around the nucleus governs the way an atom interacts with other atoms, e.g. to produce molecules. The entire list of atoms can be collated in the Periodic Table (pp. 40–1).

We now know that protons and neutrons are themselves built from smaller particles called quarks and gluons. The gluons mediate the strong nuclear force, which binds the quarks together. In total there are six types of quark, although only the lightest two of these are used in building protons and neutrons. The electrons are also part of a bigger family of particles called leptons – the muon and tau leptons are like heavier versions of the electron. The remaining leptons are three electrically neutral neutrinos.

Apart from the electromagnetic force, which causes particles with opposite electric charge to attract each other, and the strong nuclear force, there is also the weak nuclear force. This force is much weaker than the other forces, except in very high-energy interactions, and it is able to make neutrons turn into protons with the emission of an electron and a neutrino. This feature plays a key role in the nuclear fusion processes that take place in the centre of the Sun, causing it to burn.

Just as gluons mediate the strong force, the electromagnetic force is mediated by photons, which can also be regarded as particles of light. The weak nuclear force is mediated by the W and Z particles.

The Standard Model of particle physics is a very precise mathematical framework based on quantum theory that describes how all of these (i.e. the six quarks, the six leptons, and their anti-matter partners) elementary particles interact with each other through the exchange of photons, gluons and W and Z particles. The Standard Model involves one more particle, the Higgs particle, whose interactions with the other particles are responsible for their having mass. Photons and gluons do not interact with the Higgs particle and they have zero mass.

The Standard Model does not include the gravitational forces between particles, and it does not include dark matter in its list of particles. These are two of the reasons why it is generally regarded as being incomplete.

EVOLUTION OF
THE UNIVERSE

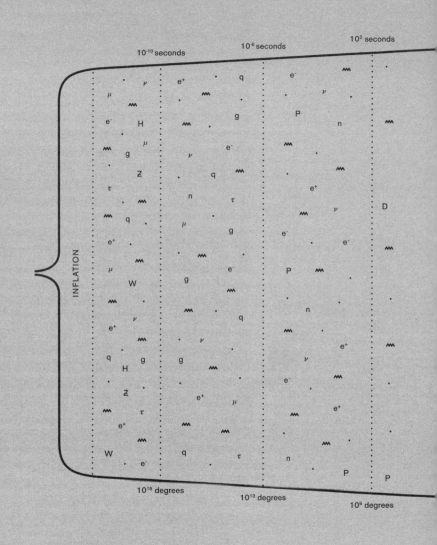

GALAXY   n NEUTRONS   q QUARKS

ν NEUTRINOS   e⁺ POSITRONS   g GLUONS

ATOMS   μ MUONS   H HIGGS BOSON

e⁻ ELECTRONS   τ TAUS   • DARK MATTER

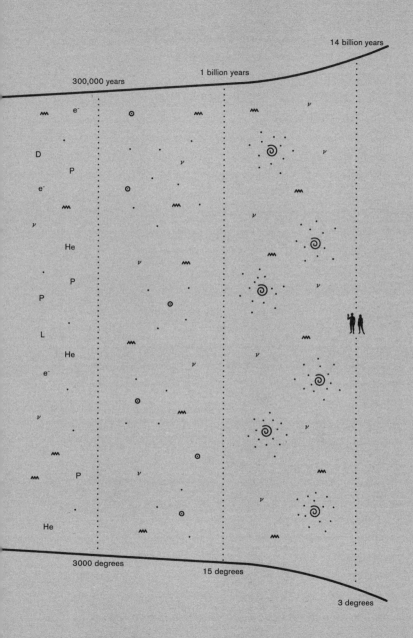

14 billion years

1 billion years

300,000 years

e⁻

D

P

e⁻

ν

He

P

P

L

He

e⁻

ν

P

He

3000 degrees

15 degrees

3 degrees

∿∿ PHOTONS

W, Z̄ WEAK FORCE CARRIERS

P, D, He, L ATOMIC NUCLEI

ILLUSTRATION CREDITS

FIGURES

2.1: data from MIT Haystack, see www.haystack.mit.edu/edu/pcr/downloads/

2.4: Data from G. Brent Dalrymple (1991), *The Age of the Earth*, Stanford University Press. Based on original data from (upper) J.-F. Minster and C. J. Allègre (1979), *Earth and Planetary Science Letters*, vol. 42, 333; (lower) S. Moorbath et al. (1977), *Nature*, vol. 270, 43

2.6: Produced using BP2000 Standard Solar Model data from J. N. Bahcall and M. H. Pinsonneault, see www.sns.ias.edu/~jnb/SNdata/sndata.html

2.9: Andrey V. Zotov and Alexander A. Saranin, see http://eng.thesaurus.rusnano.com/wiki/article939. Based on M. F. Crommie, C. P. Lutz and D. M. Eigler (1993), *Science*, vol. 262, 218

3.3: photographs by Bob Seymour

3.7: From H. Cavendish (1798), *Philosophical Transactions of the Royal Society of London*, vol. 88, 469

4.3: photographs by Kevin Kilburn

4.5: Carnegie Observatories, www.aip.org/history/exhibits/cosmology/ideas/larger-image-pages/pic-island-m31.htm

4.6: Based on the spectrum available at the NASA Extragalactic Database, see https://ned.ipac.caltech.edu/. Original reference L. C. Ho, A. V. Filippenko and W. L. Sargent (1995), *The Astrophysical Journal Supplement Series*, vol. 98, 477

4.7: Based on the spectrum in R. Walker, 'Quasar 3C273 Optical Spectrum and Determination of the Redshift, see www.ursusmajor.ch/downloads/the-spectrum-of-quasar-3c273-1.2.pdf

5.3: Based on a figure in B.P. Abbott et al. (LIGO Scientific Collaboration and Virgo Collaboration) (2016), *Physical Review Letters*, vol. 116, 061102

6.2: From E. P. Hubble (1929), *Proceedings of the National Academy of Sciences of the USA*, vol. 15, 168

6.3: Adapted from J. C. Mather et al. (1990), *The Astrophysical Journal*, vol. 354, May 10, L37

8.4: Based on http://sci.esa.int/planck/51555-planck-power-spectrum-of-temperature-fluctuations-in-the-cosmic-microwave-background/. The 2015 results of the Planck Collaboration are summarized in R. Adam et al. (2015), arXiv:1502.01582.

8.7: image by Zosia Rostomian, Lawrence Berkeley National Laboratory

8.6: Ariel Sánchez. Based on figure in A.G. Sánchez et al. (2014), *Monthly Notices of the Royal Astronomical Society*, vol. 440, 2692.

Grateful acknowledgement is given to the following for permission to reproduce or adapt material for the figures in this book:

PLATES

1: National Geographic

2: Elliot Lim and Jesse Varner, CIRES & NOAA/
NCEI, see www.ngdc.noaa.gov/mgg/image/
crustalimages.html. R. D. Müller, M. Sdrolias,
C. Gaina and W. R. Roest (2008), *Geochemistry
Geophysics Geosystems*, vol. 9, Q04006

3: Meteor Crater, Northern Arizona, USA,
www.fossweb.com/delegate/ssi-foss-ucm/
Contribution%20Folders/FOSS/multimedia/
Planetary_Science/binders/earth/earth_craters/
barringer_crater_1.html

4: Michael Goh, http://apod.nasa.gov/apod/
ap160217.html

5: hubblesite.org/NASA, http://xdf.ucolick.org/
xdf.html

6: Miodrag Sekulic, commons.wikimedia.org/wiki/
File:Bode%27s_Galaxy_Cigar_Galaxy_m81_m82.jpg

7: Based on http://dailysolar.weebly.com/
uploads/3/4/8/5/3485153/9810884.jpg

8: European Southern Observatory, www.eso.org/
public/images/potw1312a/

11: N. A. Sharp, NOAO/NSO/Kitt Peak FTS/
AURA/NSF, http://noao.edu/image_gallery/html/
im0600.html; National Optical Astronomy
Obervatory/Association of Universities for
Research in Astronomy/National Science
Foundation; calibrated axes by Larry McNish,
http://members.shaw.ca/rlmcnish/

12: S. Ossokine and A. Buonanno (Max Planck
Institute for Gravitational Physics), Simulating
eXtreme Spacetime project; scientific visualization:
D. Steinhauser (Airborne Hydro Mapping GmbH),
www.nature.com/news/gravitational-waves-how-
ligo-forged-the-path-to-victory-1.19382

14: Courtesy Caltech/MIT/LIGO Lab, www.ligo.
caltech.edu/image/ligo20150731f and www.ligo.
caltech.edu/image/ligo20150731c

13: IPAC/Caltech, by Thomas Jarrett, https://
commons.wikimedia.org/wiki/File:2MASS_LSS_
chart-NEW_Nasa.jpg

15: Chris Mihos, Case Western Reserve University/
European Southern Observatory, https://
en.wikipedia.org/wiki/Virgo_Cluster#/media/
File:ESO-M87.jpg

16: Adam Evans, https://en.wikipedia.org/wiki/
Andromeda_Galaxy#/media/File:Andromeda_
Galaxy_(with_h-alpha).jpg

17: NGC5055 (Sunflower), Gert Gottschalk;
NGC7171, The NGC/IC project; NGC1073,
Adam Block/Mount Lemmon SkyCenter/
University of Arizona; NGC3887, copyright
© 2011 The Carnegie-Irvine Galaxy Survey;
NGC3684, NGC7280, NGC0150 and UGC6533,
Sloan Digital Sky Survey; NGC0011, NGC0251
and NGC0337, Donald Pelletier; NGC4535,
Adam Block/Mount Lemmon SkyCenter/
University of Arizona; NGC4712, Robert
Gendler (www.robgendlerastropics.com), Subaru
Telescope (NAOJ), Hubble Legacy Archive,
Colour data courtesy Adam Block (http://
skycenter.arizona.edu/gallery), Bob Franke
(http://bf-astro.com) and Maurice Toet (www.
dutchdeepsky.com/index.html); NGC3668,
courtesy of Courtney Seligman

18: X-ray: NASA/CXC/CfA/M.Markevitch et al.;
Optical: NASA/STScI; Magellan/U. Arizona/D.
Clowe et al.; Lensing Map: NASA/STScI; ESO WFI;
Magellan/U. Arizona/D. Clowe et al., http://chandra.
harvard.edu/photo/2006/1e0657/index.html

19: NASA, http://hubblesite.org/gallery/album/
galaxy/cluster/pr2000007b/

20 (left): photograph by Bill Chamberlain;
(right) European Southern Observatory, https://
en.wikipedia.org/wiki/Omega_Centauri#/media/
File:Omega_Centauri_by_ESO.jpg

22: M. Blanton and Sloan Digital Sky Survey

23: NASA, http://imagine.gsfc.nasa.gov/science/
toolbox/emspectrum_observatories1.html

24: (left) Scott Kay and the Virgo Consortium
for Cosmological Simulations; (right) from J.
Schaye et al. (2015), *Monthly Notices of the
Royal Astronomical Society*, vol. 446, 521. Eagle
simulations can be obtained http://eagle.strw.
leidenuniv.nl/

25: Damien P. George, http://thecmb.org/

263

INDEX

Plate 1 The Mid-Atlantic Ridge.

0 20 40 60 80 100 120 140 160 180 200 220 240 260 280

AGE (MILLIONS OF YEARS)

Plate 2 The ages of the sea-floor rocks.

Plate 3 The Barringer Crater in Arizona. It is 1.2 km in diameter and was made by the impact of a meteorite, known as Canyon Diablo; over 10 tonnes of meteorite have been recovered from the neighbourhood. Typically, around one meteorite of size 1 metre hits the Earth every year and, fortunately, impacts from meteorites bigger than 1 km in size are expected to occur once every million years or so.

Plate 4 The Milky Way from the Australian desert.

Plate 5 The Hubble eXtreme Deep Field (XDF) image. Almost
every patch of light is a galaxy, and there are over 5500 galaxies
visible in this picture.

Plate 6 M81, Bode's Galaxy, and M82, the Cigar Galaxy. Located 12 million light years away, they are visible with binoculars. Intense star formation in M82 is caused by its interaction with M81: the two galaxies are separated by only 150,000 light years.

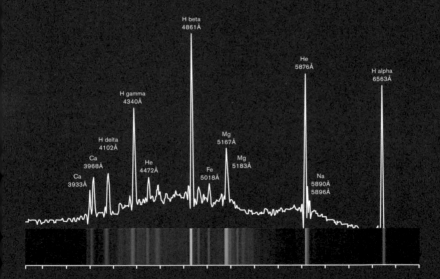

Plate 7 The solar emission spectrum, taken during
the time of a solar eclipse. Notice how the Sun has a
propensity to emit light of very specific colours. These
colours indicate which atoms are present.

Plate 8 The galaxy NGC4535 at a redshift of 0.00655.
This image was taken with the Hale 200-inch optical
reflector telescope at the Palomar Observatory located
in north San Diego County, California.

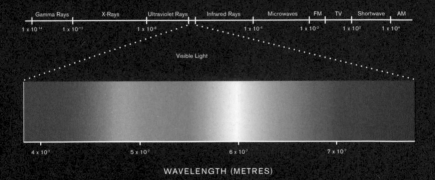

Gamma Rays   X-Rays   Ultraviolet Rays   Infrared Rays   Microwaves   FM   TV   Shortwave   AM

1×10^{-14} 1×10^{-12} 1×10^{-8} 1×10^{-4} 1×10^{-2} 1×10^{2} 1×10^{4}

Visible Light

4×10^{-7} 5×10^{-7} 6×10^{-7} 7×10^{-7}

WAVELENGTH (METRES)

Plate 9 The electromagnetic spectrum. The shortest wavelengths
are known as gamma rays, and the longest are radio waves. Visible
light has wavelengths between approximately 400 and 700 nm.
Violet and blue light have shorter wavelengths, and orange and red
have longer wavelengths.

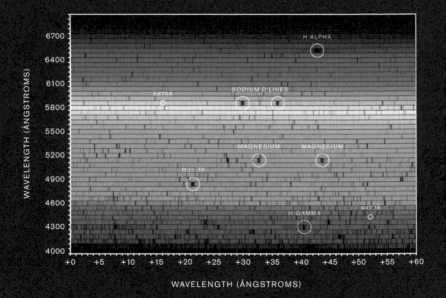

Plate 11 The solar absorption spectrum. The fingerprints of the atoms in the solar atmosphere are clearly visible, and the most prominent ones are labelled. The English chemist William Hyde Wollaston, brother of the man who gave John Michell's apparatus to Henry Cavendish, was the first person to note the appearance of these dark features in the solar spectrum, in 1802.

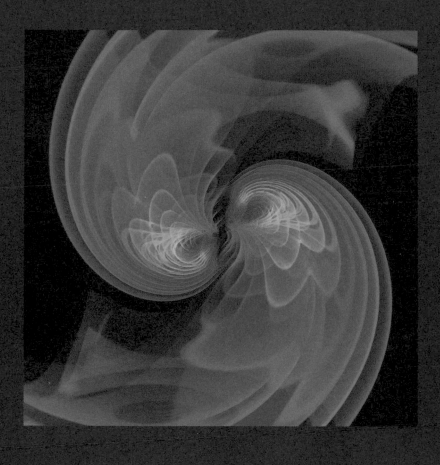

Plate 12 A simulation of a pair of colliding black holes, like the ones that produced the gravitational waves observed by the two LIGO detectors.

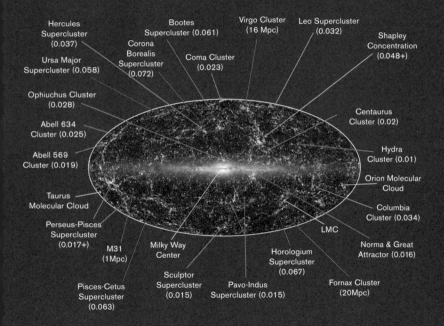

Hercules Supercluster (0.037)

Bootes Supercluster (0.061)

Virgo Cluster (16 Mpc)

Leo Supercluster (0.032)

Shapley Concentration (0.048+)

Corona Borealis Supercluster (0.072)

Coma Cluster (0.023)

Ursa Major Supercluster (0.058)

Ophiuchus Cluster (0.028)

Centaurus Cluster (0.02)

Abell 634 Cluster (0.025)

Hydra Cluster (0.01)

Abell 569 Cluster (0.019)

Orion Molecular Cloud

Taurus Molecular Cloud

Columbia Cluster (0.034)

Perseus-Pisces Supercluster (0.017+)

LMC

M31 (1Mpc)

Milky Way Center

Norma & Great Attractor (0.016)

Horologium Supercluster (0.067)

Sculptor Supercluster (0.015)

Pavo-Indus Supercluster (0.015)

Fornax Cluster (20Mpc)

Pisces-Cetus Supercluster (0.063)

Plate 13 An all-sky view of the observable Universe from the XSC catalogue of 1.5 million galaxies and the PSC catalogue of 0.5 billion stars in the Milky Way, which can be seen across the centre of the map. The galaxies are colour-coded by their redshifts.

Plate 14 The two LIGO detectors. At the top is the detector at Hanford, Washington State, and below is the detector at Livingston, Louisiana. The science is carried out by a collaboration of hundreds of people from countries across the world.

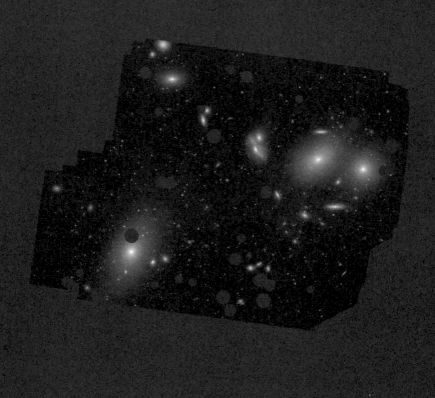

Plate 15 A deep image of the Virgo Cluster taken by
the European Southern Observatory. The dark spots
are not interesting – they are due to the removal of
bright foreground stars from the image.

Plate 16 The Andromeda galaxy –
our nearest galactic neighbour.

Plate 17 The sixteen spiral
galaxies used to measure
the rate at which space
is expanding.

NGC 1313
NGC 3668
NGC 4535
NGC 7280

NGC 0011
NGC 1073
NGC 0337
NGC 5055

NGC 0251
NGC 0150
NGC 4712
NGC 7171

NGC 3887
NGC 3684
UGC 0465
UGC 6533

Plate 18 The Bullet Cluster shows how the hot
gas (red) is displaced from the majority of the
mass inferred from gravitational lensing (blue).

Plate 19 The Hubble Space Telescope image of the Abell
2218 galaxy cluster, including the gravitationally distorted
images of more distant objects lying behind it.

Plate 20 (*left*) M3 is one of the brightest globular clusters in the
sky. At a distance of over 30 thousand light years, it is visible in
the northern hemisphere using binoculars and contains around
half a million stars; (*right*) The Omega Centauri Globular Cluster,
the largest globular cluster in the Milky Way.

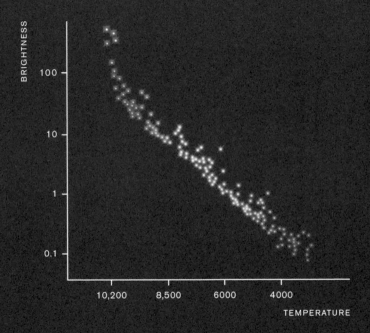

Plate 21 Herzsprung-Russell diagrams can be used
to date star clusters. The left diagram is for the
Pleiades and the right diagram is for Omega Centauri.

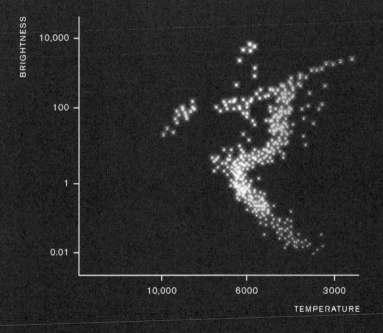

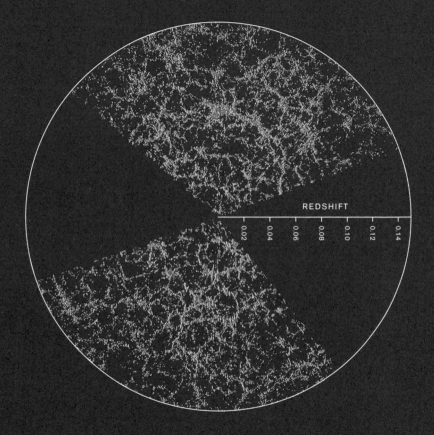

REDSHIFT

0.02 0.04 0.06 0.08 0.10 0.12 0.14

Plate 22 A map of the Universe from the Sloan Digital Sky Survey (SDSS).
The survey covers around one third of the sky and was made using a
2.5-metre optical telescope in New Mexico. Each tiny dot in the figure
corresponds to a galaxy (each one containing a few hundred billion stars).
The galaxies cluster together into wispy regions, leaving great voids.
Perhaps the greatest triumph of modern cosmology is its ability to explain,
in detail, the SDSS galaxy map and the ripples in the Cosmic Microwave
Background (CMB), shown in Figure 8.3, using the same theory.

Plate 23 The European Space Agency's Planck space telescope was launched on 14 May 2009 from the Guiana Space Centre, on an Ariane 5 rocket, and switched off on 23 October 2013. During those few years it collected data that improved on the already stunning measurements made by NASA's WMAP space telescope, which operated from 2001 until 2010. WMAP provided the first detailed measurements of the ripples in the CMB. Today, those ripples provide cosmologists with an opportunity to explore what happened at the birth of our Universe.

Plate 24 On the left is a computer simulation of how the dark matter
is spread about across a 3200 Mpc patch of simulated universe (32
Mpc deep). The brighter red regions correspond to more dark matter
and the darker regions to voids. On the right is a computer simulation
from the Eagle Project. It shows how the ordinary matter is distributed
over a smaller 100 Mpc patch of simulated universe (20 Mpc deep).
The intensity indicates the density of gas, while the colour indicates
its temperature (red is hotter). The zooms allow us to see that the
finest detail includes accurately simulated single galaxies.

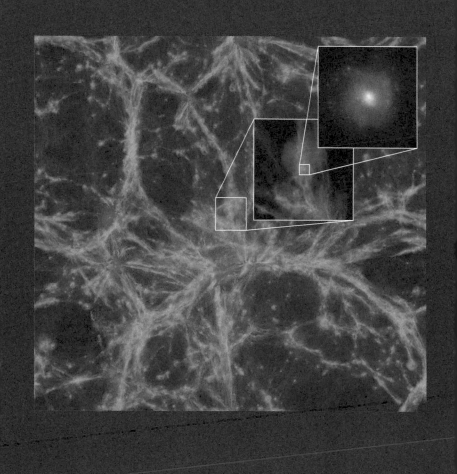

Plate 25 The Planck picture of the microwaves that bathe the Earth. These originated at the time of recombination, which is when the young hot Universe suddenly became transparent to light. This is a photograph of the Universe when it was just a few hundred thousand years old. The colours represent the temperature of the sky; the cooled, faded glow of the plasma. Hotter regions are red, and cooler regions are blue. The map shows deviations from the average temperature, which is 2.726 kelvin, or 2.726 degrees above absolute zero. The hot regions are around 100 millionths of a degree hotter than the average, and the cool regions around 100 millionths of a degree cooler. These variations in temperature are tiny and they are directly related to density fluctuations in the plasma.

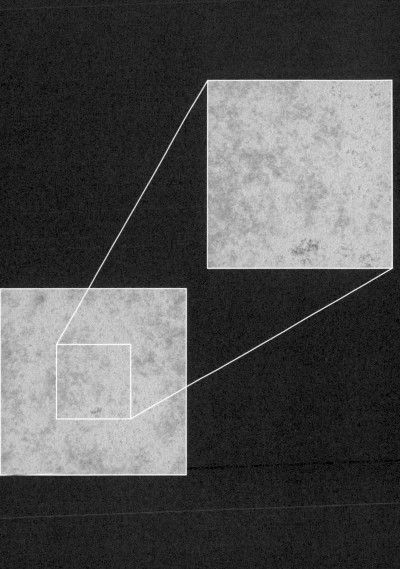

Plate 26 A snapshot of a universe during inflation.
The blue regions correspond to places where the
field is smaller. The zoomed-in portion illustrates that
the pattern is scale invariant.

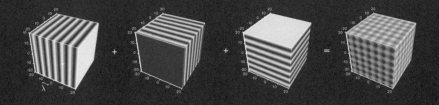

Plate 27 Adding together three plane-wave disturbances in a cube of plasma to produce the pattern of ripples indicated in the fourth cube. Absolutely any pattern of plasma disturbances can be built by adding together plane-wave disturbances in different combinations.

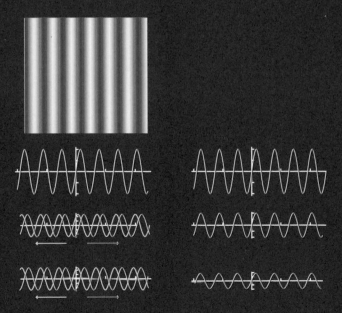

Plate 28 How standing waves are produced in the primordial plasma. The top image represents a slice through a plane wave. The graph immediately below it shows how the density varies along the wave. The graph below that is what happens a little later, when the original disturbance has started to propagate through the plasma. One half of it heads off to the left, the other half to the right. The bottom graph shows the situation still later, when the two half-waves have travelled through each other even more. The three graphs on the right are the resultant waves formed by adding together the half-waves. The resultant wave is the same shape as the original but the size is diminishing. This is a standing wave.